Experimental Agrometeorology: A Practical Manual

Latief Ahmad · Raihana Habib Kanth
Sabah Parvaze · Syed Sheraz Mahdi

Experimental Agrometeorology: A Practical Manual

 Springer

Latief Ahmad
Division of Agronomy
SKUAST-K
Srinagar, Jammu and Kashmir
India

Raihana Habib Kanth
Division of Agronomy
SKUAST-K
Srinagar, Jammu and Kashmir
India

Sabah Parvaze
Division of Agronomy
SKUAST-K
Srinagar, Jammu and Kashmir
India

Syed Sheraz Mahdi
Division of Agronomy
SKUAST-K
Shalimar, Jammu and Kashmir
India

ISBN 978-3-319-88725-8   ISBN 978-3-319-69185-5   (eBook)
https://doi.org/10.1007/978-3-319-69185-5

Printed on acid-free paper

This Springer imprint is published by Springer Nature
The registered company is Springer International Publishing AG
The registered company address is: Gewerbestrasse 11, 6330 Cham, Switzerland

Foreword

Weather and crop production are integral components of agriculture. During the era of global warming and climate change, the role of agro-meteorology in agriculture has become more important in mitigating the challenges arisen by climate change. Successful crop production depends on the prevailing weather conditions at different stages of crop growth. This manual has been prepared to enhance the understanding of undergraduate and postgraduate students regarding measurement of weather elements, their interpretation and role in agriculture and its allied subjects.

All the authors of this book deserve congratulation for bringing out this manual for the benefit of students teachers, readers and all those involved in the measurement of weather data and its application in agriculture.

Srinagar, India

Dr. Nazeer Ahmed
Vice Chancellor, SKUAST-K

Preface

Weather is the most important entity in agricultural production. Sustainable agricultural production is dependent to a large extent on the precise knowledge of the weather resources. Precise measurements of weather elements are required to understand the proper interpretation in relation to crop growth and development.

The manual "Experimental Agrometeorology" provides some of that necessary information using practical description of the "microclimate" or "crop environment". The manual contains practical assignments that deal with the measurement of weather parameters, instruments used and computation of various weather variables, crop simulation models and agro-met advisories. The manual also contains the description of agro-climatic and agro-ecological zones of India and the state of Jammu and Kashmir. The related information through glossary on the subject and tabulated Saturation Vapour Pressure, Maximum Possible Sun Shine Hours, Mean Solar Radiation and Standard Meteorological Weeks has also been included.

We hope that the manual will be helpful for undergraduate and postgraduate students of agriculture, horticulture, animal science, forestry, fisheries and other related subjects.

Nineteen chapters have been included in the manual with the aim to provide a convenient form information regarding the practices and procedures that are of the greatest importance in agricultural meteorology. The authors would be grateful to receive suggestions from readers for further improvement of this manual.

Special gratitude is expressed to Dr. Nazeer Ahmed, Hon'ble Vice Chancellor, SKUAST-K, for his able guidance and encouragement for preparation of this manual. The authors are also grateful to Prof. Sheikh Bilal Ahmad, Dean, Faculty of Agriculture, Wadura for his valuable suggestions during the preparation of the manual.

Srinagar, Jammu and Kashmir, India
August 2017

Latief Ahmad
Raihana Habib Kanth
Sabah Parvaze
Syed Sheraz Mahdi

Contents

List of Figures

List of Tables

Chapter 1
Introduction to Agricultural Meteorology

Abstract Agricultural meteorology encompasses meteorological, hydrological, pedological and biological factors that have an effect on agricultural production. It is also concerned with the interaction between agriculture and the environment. It is an inter-disciplinary science which forms a bridge between physical and biological sciences. Apart from introduction the chapter describes scope of agricultural meteorology and importance of agricultural meteorology.

Keywords Agriculture · Meteorology · Scope · Importance

1.1 Agricultural Meteorology

Agricultural meteorology encompasses meteorological, hydrological, pedological and biological factors that have an effect on agricultural production. It is also concerned with the interaction between agriculture and the environment. It is an inter-disciplinary science which forms a bridge between physical and biological sciences.

1.2 Scope of Agricultural Meteorology

Agriculture is dependent on weather more than any other aspect of human life. Increasing climatic variability coupled with climate change and increase in the incidence of extreme events has further necessitated the development and improvements in agricultural meteorology. The following applications illustrate the scope of agricultural meteorology:

Categorization of Agricultural Climate

Climatic variables like air temperature, precipitation, solar radiation, wind speed and relative humidity are important factors on which determine the growth,

© Springer International Publishing AG 2017
L. Ahmad et al., *Experimental Agrometeorology: A Practical Manual*,
https://doi.org/10.1007/978-3-319-69185-5_1

development and yield of the crop. The suitability of these parameters for increasing crop production and economical gains in a given area are assessed by agricultural meteorology.

Crop Planning for Stability in Production

Crop planning can be carried out based on water requirements of the crop, effective precipitation, soil moisture conditions, etc. in order to reduce risk of crop failure on climatic part and to stabilize yields even under hostile weather situations.

Crop Management

Crop management practices like fertilizer application, plant protection, irrigation scheduling, harvesting etc. can be carried out on the basis of specially tailored weather support. For this purpose, operational forecasts are required.

Crop Monitoring

Meteorological tools like crop growth models, remote sensing, etc. can be used to check the health and growth of crops.

Crop Modeling and Yield Climate Relationship

Suitable crop models, developed for the purpose provide information or predict the results about the growth and yield using the current and past weather data.

Research in Crop–Climate Relationship

Agricultural meteorology helps to understand relationship between crop and its climate. Thus, the complexities of plant processes in relation to its micro climate can be resolved.

Climate Extremes

Crops can be protected from climatic extremities like floods, droughts, hail, windstorms, etc. by forecasting them and warning the farmers in advance.

Analyzing Soil Moisture Stress

Climatic water balance method can be used to determine the soil moisture stress and drought. Essential protective procedures, like irrigation, mulching, application of anti-transparent, defoliation, thinning etc. can thus be undertaken to combat these situations.

Livestock Production

Livestock production is a part of agriculture. The production, growth and development of livestock is affected by the weather conditions. The weather conditions are studied in Agricultural Meteorology and breeds can be selected according to the conditions or amiable conditions can be provided for existing breeds.

Soil Formation

Climate is a foremost factor in soil formation and development as the process of soil formation depends on climatic variables like temperature, precipitation, humidity, wind etc.

1.3   Importance of Agricultural Meteorology

Climatic factors have a great impact on agricultural production. Some areas of importance of Agricultural Meteorology are listed below:

  i. Planning of cropping patterns/systems.
  ii. Selection of sowing dates for increasing crop yield.
  iii. Cost effective farm operations like ploughing, harrowing, weeding etc.
  iv. Decreasing losses of applied chemicals and fertilizers; helps to avoid fertilizer and chemical sprays when rain is forecast.
  v. Judicious irrigation to crops.
  vi. Well-organized harvesting of crops.
  vii. Reduction or elimination in outbreak of pests and diseases.
  viii. Efficient management of soils, which are formed out of weather action.
  ix. Managing weather abnormalities like cyclones, heavy rainfall, floods, drought etc. achieved by weather forecasting.

Mitigation measures such as shelterbelts against cold and heat waves, effective environmental protection, etc.

1.4   Micro-meteorology

It is the study of small scale short-lived atmospheric conditions and processes in the atmospheric surface layer. Micro-meteorology is used to understand the atmospheric processes that profoundly affect the biological activities on earth surface. Micro-meteorology controls the most important mixing and dilution processes in the atmosphere. Important topics in micro-meteorology include heat transfer and gas exchange between soil, vegetation, and/or surface water and the atmosphere caused by near-ground turbulence. Measuring these transport processes involves use of micrometeorological (or flux) towers. Variables often measured or derived include net radiation, sensible heat flux, latent heat flux, ground heat storage, and fluxes of trace gases important to the atmosphere, biosphere, and hydrosphere.

The first two meters above the ground is the disturbance zone that upsets the relatively smooth variation of climate in the synoptic layer of atmosphere-the troposphere. In micrometeorology the crop scientist is concerned with the surface layer of the ground and the air layers immediately above it. The ground surface layer which heats up by absorbing sunlight during the day is warmer than the layer above and the layer below. It, therefore, is the source of heating for these layers. After sunset the ground surface rapidly cools down as it radiates upward to the sky in the infrared region of the spectrum at the full emission rate of a black body, while it receives back from the atmosphere only the smaller amount of radiation emitted downward by the water, air, vapor, carbon dioxide and ozone in those parts of the spectrum where, air, absorption bands lie. So, very soon after sunset the cooler ground surface becomes a source of cooling and begins to get back heat from the warmer soil layers below and above it. Thus, the ground surface is an active surface because it is the source of heat during day and of cold during night.

Micrometeorology of this layer varies considerably when a crop is present. The local climate of a crop varies systematically with the crop density, the stage of growth, foliage distribution and intensity, wetness of the soil and the cultural practices. Each crop tends to develop its own characteristic microclimate. The microclimate which deviates from the neighboring open space significantly, not only provides the actual environment for the growing crop but also for pests and diseases.

Chapter 2
Agro-meteorological Observatory

Abstract A Meteorological observatory is an area where all the weather instruments and structures are installed. The chapter gives a description of a meteorological observatory, different types of agro-meteorological observatories classified by the World Meteorological Organization (WMO). The chapter gives an insight into the type of instruments required and the frequency of observations for these observatories. The site selected for the establishment of Agro-meteorological Observatory should satisfy some basic requirements. The chapter gives in detail the pre-requisites for the establishment of agro-meteorological observatory and the recommended layout of observatory. The time of observation of different parameters and the order in which the observations are taken is also discussed. Local Mean Time (LMT) is based on the average length of a solar day. It is a timekeeping method using the Sun's movements across the sky. In order to explain Local Mean Time, a few other terms need to be understood. Other terms associated with the Local Mean Time have been discussed. The chapter gives details of the Local Mean Time and the calculation of Local Mean Time of a location.

Keywords Meteorological observatory · Instruments · Layout Observations

2.1 Meteorological Observatory

An area where all the weather instruments and structures are installed is a Meteorological observatory. The instruments and structures are and exposed for measuring weather phenomena. Depending upon the type of instruments installed and frequency of observations recorded, there are 6 classes of meteorological observatories. These are Class A, B, C, D, E and F. Class A, B and C observatories are provided with both manual and self-recording instruments. The frequency of observations for Class A observatory is three times a day, twice a day for Class B observatory and once a day for Class C observatory.

© Springer International Publishing AG 2017
L. Ahmad et al., *Experimental Agrometeorology: A Practical Manual*,
https://doi.org/10.1007/978-3-319-69185-5_2

In order to record weather records, meteorological observatories are installed near aerodromes, middle of the cities and in remote areas.

2.2　Agro-meteorological Observatories and Their Classification

Agro-meteorological observatories are established for observation and recording of meteorological as well as biological parameters of crops using different instruments. According to World Meteorological Organization (WMO 2003), every agro-meteorological station belongs to one of the following classes:

Principal Agro-meteorological Station

A Principal Agro-Meteorological Station provides detailed simultaneous meteorological and biological information and its established where research in agricultural meteorology is carried out.

Requirements:

a. **Essential instruments**

1. Maximum and minimum thermometers.
2. Wet and dry bulb thermometers.
3. Soil thermometers.
4. Grass minimum thermometer.
5. Rain gauge (ordinary and self- recording).
6. Wind vane and anemometer.
7. USWB. Open pan evaporimeter.
8. Sunshine recorder.
9. Assmann psychrometer.
10. Dew gauge.
11. Thermo hygrograph.
12. Soil moisture equipment.
13. Solar radiation instruments.

b. **Optional instruments**

1. Lysimeter.
2. Thermopile sensing elements for short and long wave net radiation.
3. Potentiometer.
4. Micro voltmeter.

Ordinary Agro-meteorological Station

An ordinary agricultural meteorological station provides, on a routine basis, simultaneous meteorological and biological information and may be equipped to assist in research into specific problems.

Requirements:

a. **Essential instruments**

1. Maximum and minimum thermometers.
2. Wet and dry bulb thermometers.
3. Soil thermometers.
4. Grass minimum thermometer.
5. Rain gauge (ordinary).
6. USWB open pan evaporimeter.
7. Assmann Psychrometer.

b. **Optional instruments**

1. Sunshine recorder.
2. Dew gauge.
3. Self-recording rain gauge.
4. Thermo-hygrograph.

An Auxiliary Agro-meteorological Station

An Auxiliary Agro-Meteorological Station provides meteorological and biological information. The meteorological information may include such items as soil temperature, soil moisture, potential evapotranspiration, duration of vegetative wetting, and detailed measurements in the very lowest layer of the atmosphere. The biological information may cover phenology, onset and spread of plant diseases, and so forth.

Requirements:

a. **Essential instruments**

1. Maximum and minimum thermometers.
2. Dry bulb and wet bulb thermometers.
3. Ordinary rain gauge.

b. **Optional instruments**

1. Wind vane and anemometer.
2. Dew gauge.

Agricultural Meteorological Station for Specific Purposes

This is a station set up temporarily or permanently for the observation of one or several variables and/or specified phenomena.

2.3 Site Selection for Agro-meteorological Observatory

The site selected for the establishment of Agro-meteorological Observatory should satisfy the following basic requirements.

1. The site should be representative of the crop-soil-climate conditions of the area. Representatively of a measurement is the degree to which it describes reliably the value of some parameter (for instance, humidity or wind speed) at a specified space scale for a specified purpose (WMO 2001).
2. It should be located at the center of the farm.
3. The site should be free from water logging.
4. It should have easy accessibility during the rainy season.
5. The site should be away from any permanent irrigation sources and tall structures like buildings, hillocks and trees.
6. The site should not have extreme topography and it should be well exposed.
7. The site of a weather station should be fairly level and under no circumstances should it lie on concrete, asphalt, or crushed rock. Wherever the local climate and soil do not permit a grass cover, the ground should have natural cover common to the area, to the extent possible.
8. Obstructions such as trees, shrubs and buildings should not be too close to the instruments.
9. Sunshine and radiation measurements can be taken only in the absence of shadow during the greater part of the day; brief periods of shadows near sunrise and/or sunset may be unavoidable.
10. Wind should not be measured at a proximity to obstructions that is less than ten times their height.
11. Tree drip into rain gauges should not be allowed to occur.

2.4 Recommended Layout of Observatory

The dimensions for an observatory are a length of 55 m and width of 36 m and the longer side running South–North. The ground plan for an agro-meteorological observatory is given in Fig. 2.1. The periphery should be fenced with barbed wires to prevent cattle trespass. There should be a gate at appropriate site. All tall instruments should be installed at the northern side of the observatory to avoid shade effect.

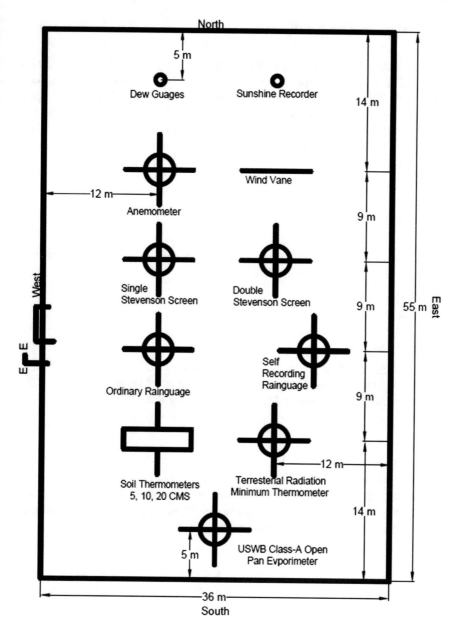

Fig. 2.1 Layout of agro-meteorological observatory

2.5 Time of Observation

Regular surface observations are taken at 0530, 0830, 1130, 1430 1730, 2030 and 2330 h IST at Class I observatories. Morning and afternoon observations refer to 0830 and 1730 h IST observations, respectively. But, agro meteorological observations are taken at 0700 and 1400 h LMT, which when converted into IST. Rainfall and evaporation observations are taken at 08.30 h. Indian Standard Time (IST) and 14.00 h LMT. The setting of automatic instruments like thermograph, hydrograph, evaporigraph and barographs etc. are done at 08.30 h IST.

2.6 Order of Observations

The instruments at the observatory should be read in the following order commencing from 10 min preceding the hour

(1) Wind instruments,
(2) Rain gauge,
(3) Thermometers and
(4) Barometer.

Non-instrumental observations (e.g. clouds, visibility etc.) should be taken in the interval of 5 min between the first and second readings of the anemometer or if that is not possible, before commencing the instrumental observations.

2.7 Local Mean Time

Local Mean Time (LMT) is based on the average length of a solar day. It is a timekeeping method using the Sun's movements across the sky. In order to explain Local Mean Time, a few other terms need to be understood.

Apparent Solar Time/True Solar Time

A slight variation in the length of solar days is caused due to the rotation of the earth. This implies that the speed of true solar time is not constant. True or apparent solar times shown by means of a sundial.

Mean Solar Time

The length of a mean or average solar day, i.e. 24 h is the Mean Solar. It moves at a constant speed. Local Mean Time is the Mean Solar Time for a specific location on Earth. Places sharing same longitude have the same Local Mean Time.

2.8 Calculation of Local Mean Time

The Local Mean Time corresponding to Indian Standard Time (IST) varies from place to place depending upon longitude of a place. The longitude $82° 30'$ E passing through Allahabad is called Indian Standard Time longitude. Here the IST and LMT are same. To get IST of a corresponding LMT of a place the following formula can be used.

$$IST = LMT + 4(\lambda_S - \lambda_L)$$

where,

LMT   7:00 h.
λ_S Standard time longitude i.e., $82° 30'$E for India passing through Allahabad.
λ_L Longitude of the station for which local time is calculated.

For traversing the distance between two longitudes, the earth takes nearly 4 min. So the 4 is taken as the multiplication factor for the difference.

Example

The longitude of Srinagar is $74.8°$. The time of observation (i.e. IST) at Srinagar corresponding to 07:00 and 14:00 h LMT can be calculated as:

$$IST = 07:00 + 4(82.5 - 74.8)$$
$$= 07:00 + 00:04 \, (7.7)$$
$$= 07:00 + 00:31$$
$$07:31 \, (\text{approx.})$$

and,

$$IST = 14:00 + 4(82.5 - 74.8)$$
$$= 14:00 + 00:04 \, (7.7)$$
$$= 14:00 + 00:31$$
$$14:31 \, (\text{approx.})$$

Chapter 3
Measurement of Temperature

Abstract Temperature symbolizes the thermodynamic condition of a body. The direction of the net flow of heat between two bodies determines the value of temperature. For meteorological purposes, temperatures are measured for a number of media. The most common variable measured is air temperature (at various heights), ground, soil, grass minimum and seawater temperature. The chapter describes measurement of temperature of air and soil required for agro-meteorology. Various instruments used for measurement of air and soil temperature have been discussed.

Keywords Temperature · Stevenson's screen · Units · Temperature measurement

3.1 Temperature

Temperature is defined as "a physical quantity characterizing the mean random motion of molecules in a physical body" (WMO 1992). Temperature symbolizes the thermodynamic condition of a body. The direction of the net flow of heat between two bodies determines the value of temperature. For meteorological purposes, temperatures are measured for a number of media. The most common variable measured is air temperature (at various heights), ground, soil, grass minimum and seawater temperature.

WMO (1992) defines air temperature as "the temperature indicated by a thermometer exposed to the air in a place sheltered from direct solar radiation". Air temperature has a significant influence on crop growth and development. The temperature range prevailing during the crop growth season determines the productivity of the crops. Temperature also influences the water requirements of crops.

3.2 Units of Temperature

The temperature is usually given in degrees Celsius (°C) or degrees Fahrenheit (°F). The SI unit of temperature is Kelvin (K).

3.3 Measurement of Temperature

The following temperature measurements are required from the agro-meteorological point of view:

i.   The air near the Earth's surface;
ii. The surface of the ground;
iii. The soil at various depths;
iv. The surface levels of the sea and lakes;
v.   The upper air.

3.4 Measurement of Air Temperature

Liquid-in-glass thermometers are commonly used for routine observations of air temperature, including maximum, minimum, dry bulb and wet-bulb temperatures. The liquid used in a thermometer depends on the required temperature range; mercury is generally used for temperatures above its freezing point (-38.3 °C), while ethyl alcohol or other pure organic liquids are used for lower temperatures. Air temperature observations are usually carried out by means of thermometers exposed in a standard screen known as Stevenson's Screen.

Stevenson's Screen

This Screen was designed by Thomas Stevenson in 1866. It is a specialized shelter to accommodate the four thermometers i.e., maximum, minimum, dry bulb and wet bulb thermometers for recording of air temperature (Fig. 3.1). Stevenson's screen is a wooden rectangular box of the following dimensions: Length 56 cm, width 30 cm and height 40 cm with a double-layered roof and louvered sides. The screen is painted white and is mounted on four wooden supports. The support of the screen is placed at a height of 4 ft. (1.22 m) above the ground. The screen is set up with its door facing north side (opening downward) so that minimum sunlight would enter while the observer is reading the instruments.

The screen is meant to protect the thermometers from direct heating from ground and neighboring objects and from losing heat by radiation at night. The double layered roof and louvered sides protects the instruments from rain and snow. It also allows free air circulation.

The maximum and minimum thermometers are laid in horizontal positions on the upper and lower wooden brackets respectively and rest at an angle of 2° to horizontal. The dry bulb and wet bulb thermometers are kept vertical on the wooden bracket on the left and right hand sides respectively.

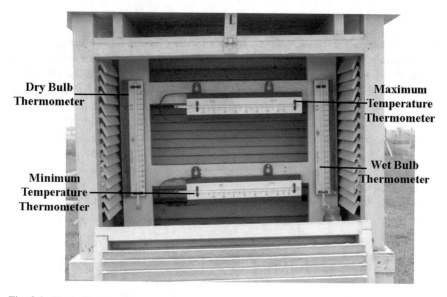

Fig. 3.1 Single Stevenson screen

Precautions

1. Take the temperature reading as quick as possible so that it is not affected by presence of the observer.
2. Avoid parallax error while reading the thermometers.
3. Do not keep the door of the thermometer screen open for a longer time.
4. Use distilled water or rain water for the wet bulb thermometer and keep the container clean.
5. Keep muslin cloth and threads clean and free from dust and grease. Change them every week.
6. Keep the water container away from dry bulb thermometer and do not keep directly below the wet bulb thermometer.
7. Correction factors should be applied for correction the air temperature.

Maximum Temperature Thermometer

The maximum temperature thermometer is a mercury-in-glass thermometer provided with a constriction in the capillary of the glass tube below the lowest graduation of the scale. The constriction permits the mercury to be pushed forward by rising temperature but restricts it being drawn back as the temperature falls. It stands at that level in the capillary, so that maximum temperature can be recorded at a later time. The range of the thermometer is −35 to +55 °C. The thermometer is set at 07.00 h LMT. The reading of the maximum thermometer after setting should agree with that of dry bulb thermometer within 0.3 °C. The thermometer is set by brisk shaking.

Minimum Temperature Thermometer

The minimum or the lowest temperature of air during last 24 h is measured with the help of minimum temperature. It is a spirit or alcohol-in-glass thermometer having a range of −40 to +50 °C. The thermometer has a light narrow index in the stem. This index is kept inside the spirit column by the surface tension of the meniscus. Reading is taken from the end of the index which is farthest from the bulb.

As the temperature falls, the alcohol contracts and end of alcohol column in stem moves towards the bulb dragging the index along with it by the surface tension of liquid. If subsequently temperature increases the alcohol flows freely past the index without displacing it. Thus, the position of the end of the index farthest from the bulb indicates the lowest temperature reached since the thermometer was last set. It is set at 14.00 h LMT by tilting the bulb upwards. After setting, the reading of the minimum thermometer should agree with that of dry bulb thermometer within 0.6 °C.

Dry Bulb Thermometer

Dry Bulb Temperature is the ambient air temperature and is measured with a mercury-in-glass thermometer. It is an ordinary thermometer with a temperature range of −35 to +55 °C. Dry Bulb temperature is called so because air temperature is indicated by a thermometer not affected by the moisture of the air.

Wet Bulb Thermometer

Wet Bulb temperature is measured by using a Wet Bulb Thermometer. It is an ordinary thermometer with the bulb wrapped in wet muslin. The bulb acts as an evaporating surface. The bulb is kept continuously wet by providing water by means of four strands of cotton thread dipped into a small water container with distilled water or rain water. The adiabatic evaporation of water from the thermometer bulb results in the cooling effect and thus the "wet bulb temperature" is lower than the "dry bulb temperature" in the air.

The rate of evaporation from the muslin cloth on the bulb and the temperature difference between the dry bulb and wet bulb depends on the humidity of the air. The evaporation from the wet muslin is reduced when air contains more water vapor. When the air is saturated, both the dry bulb and wet bulb temperatures would be same. This temperature is used to find out dew point temperature, vapor pressure and humidity.

Thermograph

A mechanical thermograph is an automatic temperature recording instrument having a rotating chart mechanism which provides a continuous plot of temperature against time. There are two types of mechanical thermographs:

i. **Bimetallic thermograph**

In bimetallic thermographs, the movement of the recording pen is controlled by the change in curvature of a bimetallic strip or helix, one end of which is rigidly fixed to an arm attached to the frame. The bimetallic element should be adequately protected from corrosion; this is best done by heavy copper, nickel or chromium plating, although a coat of lacquer may be adequate in some climates.

ii. Bourdon tube thermograph

The general arrangement is similar to that of the bimetallic type but its temperature-sensitive element is in the form of a curved metal tube of flat, elliptical section, filled with alcohol. The Bourdon tube is less sensitive than the bimetallic element and usually requires a multiplying level mechanism to give sufficient scale value.

Dew Point Temperature

The Dew Point temperature is the temperature to which the air should be cooled in order just to cause condensation. It is also defined as the temperature at which the saturation vapor pressure is equal to the pressure of the vapor present in the air. The Dew Point temperature at a particular Dry Bulb and Wet Bulb temperature set is also estimated by means of hygrometric tables provided by IMD.

3.5 Measurement of Soil Temperature

Soil temperature in simple words is the measurement of the warmth in the soil. Soil temperature is an important factor for plant growth and development. Soil temperature affects the water movement within the soil and plays a vital role in germination of seeds and root system development. The activity of the soil micro-flora, decomposition of organic material and absorption capacity of roots depend on the soil temperature. An increase in soil temperature results in the increased solubility of some major salts.

Soil Thermometer

A soil thermometer is a mercury-in-glass thermometer. The bulb of this thermometer is bent at an angle of 120° while the stem is straight. The range of the thermometer is −20 to +60 °C. The site for such measurements should be a level plot of size 180 cm × 120 cm in the observatory and typical of the surrounding soil for which information is required. The standard depths for measurement of soil temperature are 5, 10 and 20 below the surface. Additional depths may however be included. The thermometers are placed 45 cm apart at an inclined depth of 5.8, 11.6 and 23.2 cm to ensure a vertical depth of 5, 10, 15 and 20 cm respectively (Fig. 3.2). In places where ground is covered with snow, it is desirable to measure the temperature of snow cover as well.

Installation

i. Minimum soil layer should be disturbed during installation of thermometers inside the soil.
ii. The bulb of the thermometer should rest in good contact with the firm undisturbed soil.

Fig. 3.2 Soil thermometers

  iii. While digging the soil for installation, the soil should be removed layer wise and identically layers put in corresponding previous position.

  iv. The zero mark of the thermometer should be at ground level.

Recording Observations

The soil temperature readings are taken read daily at 07.00 and 14.00 h LMT. The following points should be kept in mind while taking the observations:

  i. The line joining the observer's eye and top of the mercury column should be at right angle to the instrument.

  ii. The observer should stand at some distance from the instrument so as not to shadow the ground surface in the vicinity of the instrument.

  iii. Reading should be taken correct to 0.10 °C.

Grass Minimum Thermometer

The grass minimum temperature is the lowest temperature reached overnight by a thermometer freely exposed to the sky just above short grass. The grass minimum temperature is measured with a minimum thermometer as described in Section "Minimum Temperature Thermometer". The stem of this thermometer is however encased in a glass jacket in order to protect seal marks on the stem from dew, rain and prevent the stem from cooling rapidly. Grass minimum temperature indicates the chance of occurrence of ground frosts. A ground frost is likely to occur when the instrument records a temperature below 0 °C (Fig. 3.3).

Fig. 3.3 Grass minimum thermometer

Thermometers are mounted on suitable Y-shaped supports so that they are inclined at an angle of about 2° from the horizontal position, with the bulb being lower than the stem. The reading is taken before sunrise. After reading, the instrument is kept indoors to avoid direct exposure to solar radiation. The grass minimum temperature remains lower than the air temperature at screen level.

3.6   Computation of Soil Heat Flux

Soil heat flux is measured by means of Soil Heat Flux plates (Fig. 3.4) placed at suitable depths in the soil usually 8 cm. Temperature gradient across the plates is measured by means of a thermopile. Each plate is individually calibrated to output flux (Fig. 3.5).

Fig. 3.4 Heat flux plates

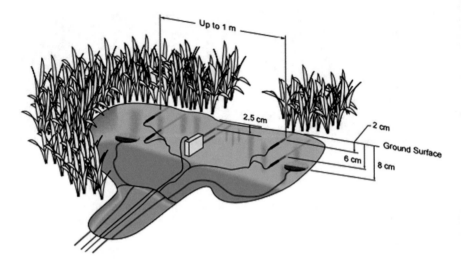

Fig. 3.5 Placement of heat plates and sensors for measurement of soil heat flux

The measured flux G(d) at a depth 'd' is added to the energy stored in the layer above the heat flux plates to calculate the surface heat flux. The specific heat of the soil and the change in soil temperature, ΔT_s, over the output interval, t, are required to calculate the stored energy and thus the surface flux. The steps for calculating the surface flux are given as:

i. The heat capacity of the soil is calculated by adding the specific heat of the dry soil to that of the soil water. The values used for specific heat of dry soil and water are on a mass basis. The heat capacity of the moist is given by:

$$C_s = \rho_b(C_d + \theta_m C_w) = \rho_b C_d + \theta_v \rho_w C_w$$

$$\theta_m = \frac{\rho_w}{\rho_b}\theta_v$$

where,

C_s Heat capacity of moist soil,
ρ_b Soil bulk density,
C_d Heat capacity of a dry mineral soil,
θ_m Soil water content on a mass basis,
θ_v Soil water content on a volume basis,
C_w Heat capacity of water,
ρ_w Density of water.

ii. The soil storage term is calculated as:

$$S_{soil} = \frac{\Delta T_s C_s d}{t}$$

iii. The surface flux (G_s) is then calculated as:

$$G_s = S_{soil} + G(d)$$

Chapter 4
Measurement of Humidity

Abstract Humidity measurements at the Earth's surface are required for meteo-rological analysis and forecasting, for climate studies, and for many special applications in hydrology, agriculture, aeronautical services and environmental studies, in general. They are particularly important because of their relevance to the changes of state of water in the atmosphere. The chapter describes instruments used for measurement of Relative Humidity using psychrometers and hygrometer. As section on estimation of relative humidity from temperature data has also been included.

Keywords Humidity · Measurement · Psychrometer · Estimation

4.1 Introduction

Humidity measurements at the Earth's surface are required for meteorological analysis and forecasting, for climate studies, and for many special applications in hydrology, agriculture, aeronautical services and environmental studies, in general. They are particularly important because of their relevance to the changes of state of water in the atmosphere. The measurement of atmospheric humidity, and often its continuous recording, is an important requirement in agro-meteorology. This chapter deals with the measurement of humidity at or near the Earth's surface. Some terms used in this chapter are defined below:

i. **Mixing ratio r**: The ratio between the mass of water vapor and the mass of dry air;
ii. **Absolute humidity**: The weight of water vapor present per unit volume of air;
iii. **Specific humidity q**: The ratio between the mass of water vapor and the mass of moist air;
iv. **Dew point temperature Td**: The temperature at which moist air saturated with respect to water at a given pressure has a saturation mixing ratio equal to the given mixing ratio;

© Springer International Publishing AG 2017
L. Ahmad et al., *Experimental Agrometeorology: A Practical Manual*,
https://doi.org/10.1007/978-3-319-69185-5_4

    v. **Relative humidity U**: The ratio in percent of the observed vapor pressure to the saturation vapor pressure with respect to water at the same temperature and pressure;

    vi. **Vapor pressure e**: The partial pressure of water vapor in air;

    vii. **Saturation vapor pressures e_w and e_i**: Vapor pressures in air in equilibrium with the surface of water and ice, respectively.

4.2 Units of Measurement

The units and symbols generally used for expressing the most commonly used quantities associated with water vapor in the atmosphere are listed in Table 4.1.

4.3 Measurement of Relative Humidity

The instruments required for measurement of relative humidity are:

1. Simple or stationary psychrometer.
2. Assmann psychrometer.
3. Whirling psychrometer.
4. Hair hygrometer.

Simple or Stationary Psychrometer

A simple psychrometer is a set of dry bulb and wet bulb thermometers of identical form and size exposed in a Stevenson's Screen. The dew point temperature and relative humidity are then estimated from the hygrometric tables.

Assman Psychrometer

A pair of vertically fixed dry and wet bulb thermometers with cylindrical bulbs form this psychrometer. The bulbs are protected from external radiation by means

Table 4.1 Units and symbols of some of the commonly used quantities associated with water vapor in the atmosphere

Quantity	Symbol	Unit
Mixing ratio	r	$kg\ kg^{-1}$
Absolute humidity	Q	$kg\ kg^{-1}$
Specific humidity	q	$kg\ kg^{-1}$
Vapor pressure in air	e	Pa
Saturation vapor pressures	e_w, e_i	Pa
Pressure	P	Pa
Relative humidity	U	percent

Fig. 4.1 Assman
psychrometer

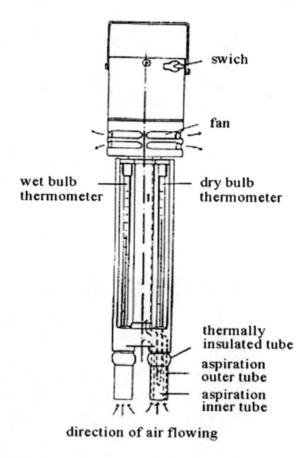

swich

fan

wet bulb
thermometer

dry bulb
thermometer

thermally
insulated tube

aspiration
outer tube

aspiration
inner tube

direction of air flowing

of two polished coaxial tubes (Fig. 4.1). The temperature and RH of both open and
inside crops are measured by this psychrometer.

Whirling Psychrometer

It consists of a set of two thermometers attached horizontally to a rectangular
wooden frame which can be rotated with a handle (Fig. 4.2). The psychrometer is
designed to measure temperature and Relative Humidity of both inside and outside
crops.

Hair Hygrometer

The hair hygrometer uses the characteristic of the hair that its length expands or
shrinks response to the relative humidity. The length of human hair from which
liquid are removed increases by 2–2.5% when relative humidity changes by 0–
100%. The hair hygrograph is a hair hygrometer to which a clock-driven drum is
installed to record humidity on a recording chart (Fig. 4.3).

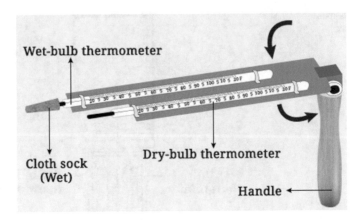

Fig. 4.2 Whirling psychrometer

Fig. 4.3 Hair hygrometer

4.4   Estimation of Relative Humidity from Temperature

Relative humidity is most conveniently estimated using Dry Bulb and Wet Bulb temperatures. IMD uses August's modification of Regnault's formula (temperature in °C instead of °F) for calculation of vapor pressure and thus the hygrometric tables. The equation is given as follows:

For temperatures of wet bulb below 0 °C:

$$X = F' - \frac{0.480\,(T - T')}{671 - T'} \times p$$

For temperatures of wet bulb above 0 °C:

$$X = F' - \frac{0.480\,(T - T')}{610 - T'} \times p$$

where,

X   Pressure of vapor present in the air;
F'   Saturation vapor pressure at temperature of the Wet Bulb;
T   Temperature of the Dry Bulb in °C;
T'   Temperature of the Wet Bulb in °C;
p   Pressure of air.

The hygrometric tables calculated from these equations are applicable only when the thermometers are in a Stevenson Screen or exposed to a gentle breeze at the time of observation.

The RH is calculated as:

$$U = \frac{X}{F} \times 100$$

where,

U   Relative Humidity;
F   Saturation vapor pressure at temperature of the Dry Bulb.

Various tables associated with the estimation of dew point temperature and RH are given in Appendix A.

Chapter 5
Measurement of Wind

Abstract Wind plays an important role in crop evapotranspiration and thus determines crop water use. The measurement of wind is thus necessary for studying the crop growth. The chapter deals with the measurement of wind speed and direction and the instruments used for the same. The units used for expression of wind speed and their correlation have been given. The chapter also introduces wind rose and Beaufort Scale.

Keywords Wind · Measurement · Observations · Units

5.1 Introduction

Wind refers to the horizontal movement of the earth in response to differences in pressure. Wind is a three-dimensional vector that has the directions of north, south, east and west in addition to vertical components and magnitude (i.e., wind speed). For most operational meteorological purposes, the vertical component is ignored, surface wind is practically considered as a two-dimensional vector.

Wind plays an important role in crop evapotranspiration and thus determines crop water use. The measurement of wind is thus necessary for studying the crop growth.

5.2 Measurement of Wind Direction

The instrument used for the measurement of wind is wind vane. The direction from which wind blows is known as the windward side and towards which it bows is known as leeward side.

© Springer International Publishing AG 2017
L. Ahmad et al., *Experimental Agrometeorology: A Practical Manual,*
https://doi.org/10.1007/978-3-319-69185-5_5

Fig. 5.1 Wind vane

Wind Vane

Wind vane consists of a brass-arm, mounted on ball bearing to a vertical axis (Fig. 5.1). To one side of the brass arm there is an arrowhead and on another side there are two flat vanes forming an acute angle (about 20°). The ball bearings are in bearing house where there is an oil hole. The screw below the bearing are in bearing house are tightened to a brass covering known as brass sleeve. Below the wind vane there are 4 direction arms fixed to the vertical axis by means of a brass boss. In between the direction arms there are corner indicators. The direction arms and corner indicators are tightened to the brass boss by means of knots known as check knots. The direction arms are levelled with N, S, E and W. The vertical axis is erected by means of an iron-stand.

Units

Wind direction can be expressed in two ways:

1. By direction (Sixteen point of a compass).
2. By degree (from north, measured in clockwise as N, E, S, W means 360°, 90°, 180° and 270° respectively).

Observation

Watch the wind vane for a few minutes and identify the direction nearest to the sixteen point of the compass (Fig. 5.2).

Fig. 5.2 Wind direction

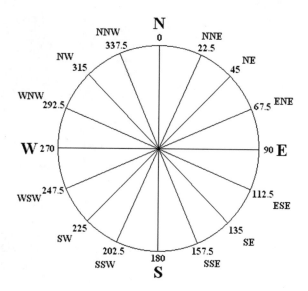

5.3   Measurement of Wind Speed

Wind speed is measured by means of an anemometer. There are two types of
rotating anemometers: the cup anemometer, which has three or four cup wheels
attached to the rotating axis, and the propeller anemometer, which has propeller
blades. Both types rely on the principle that the revolution speed of the cup or
propeller rotor is proportional to the wind speed.

Cup Anemometers

A cup anemometer has three or four cups mounted symmetrically around a free-
wheeling vertical axis (Fig. 5.3). The difference in the wind pressure between the
concave side and the convex side of the cup causes it to turn in the direction from
the convex side to the concave side of next cup. The revolution speed is proportional to the wind speed irrespective of wind direction. Wind speed signals are
generated with either a generator or a pulse generator.

The cups were conventionally made of brass but in recent years, cups made of
light alloy or carbon fiber thermo-plastic have become the common, allowing
significant reductions in weight. Beads are set at the edges of the cups to add
rigidity and deformation resistance. They also help the cup to avoid the effects of
turbulence, allowing the stable measurement of a wide range of wind speeds.

The anemometer is installed on a metal pipe, which is fixed on a wooden post.
The height from the center of the anemometer cups is 10 ft. above the ground level.

Propeller Anemometers

A propeller anemometer has a sensor with a streamlined body and a vertical tail to
detect wind direction and a sensor in the form of a propeller to measure wind speed

Fig. 5.3 Cup anemometer

Fig. 5.4 Propeller
anemometer

integrated into a single structure (Fig. 5.4). It measures wind direction and wind
speed, and can indicate/record the instantaneous wind direction and wind speed in
remote locations. It also measures the average wind speed using wind-passage
contacts or by calculating the number of optical pulses.

Units

A number of different units are used to indicate wind speed. Some of the common
units of measurement are

Table 5.1 Conversion of
units of wind speed

kt	m/s	km/h	mph	ft./s
1	0.515	1.853	1.152	1.689

  i. meters per second (m/s),
 ii. kilometers per hour (km/h),
iii. miles per hour (mph),
iv. feet per second (ft./s) and
 v. knots (kt).

The conversion of these units is given in Table 5.1:

Observation

The reading of wind speed is obtained from the cyclometer dial having a range of
0–9999. The four black figures give whole km and the red figure to the right gives
tenth of km. Wind speed is calculated as follows:

1. Note down two reading from anemometer at an interval of 3 min. Multiply the
   difference by 20 to get wind speed at the time of observation in km/h.
2. Subtract the anemometer reading at 07.00 h LMT of the previous day from that
   at 07.00 h LMT of the observation day and divide the difference by 24 to get the
   mean daily wind speed for the observation day in km per hour.

5.4 The Beaufort Wind Scale

If a measuring instrument becomes faulty or is not available, wind can be estimated
by visual means such as observing smoke as a guide to wind speed and using the
Beaufort Scale (Table 5.2). The Beaufort scale was first developed by Admiral
Francis Beaufort in 1806. It emphasizes more on the observed effect of the wind,
rather than the actual wind speed.

5.5 Wind Rose

Wind direction is normally defined by a wind rose. Wind rose is a graphic display
of the distribution of wind direction at a location during a defined period. The
characteristic patterns can be presented in either tabular or graphic forms. A wind
rose is a set of wind statistics that describes the frequency, direction, force, and
speed (Fig. 5.5). In this plot the average wind direction is shown as one of the
sixteen compass points, each separated by 22.5′ measured from true north. The
length of the bar for a direction indicates the percent of time the wind came from
that direction. Since the direction is constantly changing, the time percentage for a

Table 5.2 Beaufort wind scale

Beaufort scale number and description	Wind speed equivalent at a standard height of 10 m above open flat ground				Specifications for estimating speed over land
	(kt)	(m/s)	(km/h)	(mph)	
0 Calm	<1	0–0.2	<1	<1	Calm; smoke rises vertically
1 Light air	1–3	0.3–1.5	1–5	1–3	Direction of wind shown by smoke-drift but not by wind vanes
2 Light breeze	4–6	1.6–3.3	6–11	4–7	Wind felt on face, leaves rustle; ordinary vanes moved by wind
3 Gentle breeze	7–10	3.4–5.4	12–19	8–12	Leaves and small twigs in constant motion, wind extends light flags
4 Moderate breeze	11–16	5.5–7.9	20–28	13–18	Raises dust and loose paper, small branches are moved
5 Fresh breeze	17–21	8.0–10.7	29–38	19–24	Small trees in leaf begin to sway, crested wavelets form on inland waters
6 Strong breeze	22–27	10.8–13.8	39–49	25–31	Large branches in motion, whistling heard in telegraph wires; umbrellas used with difficulty
7 Near gale	28–33	13.9–17.1	50–61	32–38	Whole trees in motion, inconvenience felt when walking against the wind
8 Gale	34–40	17.2–20.7	62–74	39–46	Breaks twigs off trees, generally impedes progress
9 Strong gale	41–47	20.8–24.4	75–88	47–54	Slight structural damage occurs (chimney-ports and slates removed)
10 Storm	48–55	24.5–28.4	89–102	55–63	Seldom experienced inland, trees uprooted, considerable structural damage occurs
11 Violent storm	56–63	28.5–32.6	103–117	64–72	Very rarely experienced; accompanied by widespread damage
12 Hurricane	64 and over	32.7 and over	118 and over	73 and over	–

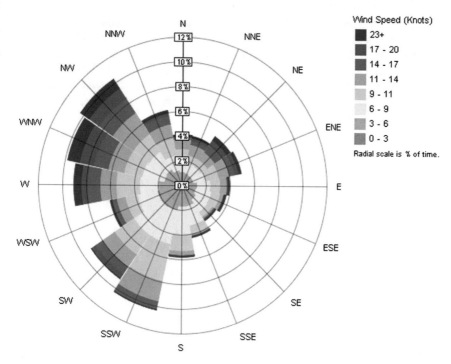

Fig. 5.5 Wind rose showing direction and velocity frequencies

compass point includes those times for wind direction at 11.25° on either side of the point. The percentage of time for a velocity is shown by the thickness of the direction bar.

Preparation of Wind Rose

i. A circle is first divided into 8, 12, or 16 sectors representing the directions of wind flow.

ii. The percentage of time the wind blows from each direction is determined from wind direction data. Table 5.3 shows a sample frequency data table for preparation of wind rose.

Table 5.3 Sample frequency data table for preparation of wind rose

Direction	Frequency	Percentage (%)
N	165	22
NE	65	14
E	142	5
SE	17	3
S	65	9
SW	59	11
W	38	9
NW	150	27
Total	701	100

iii. The data is then plotted on a circular graph as a line emanating from the center of the circle. The length of the line is scaled to the percentage obtained from the data, pointing in the given direction.

iv. The wind speed can be included in wind rose diagram. For this purpose, the percent of time that the wind is blowing at given speeds toward each direction needs to be determined. This data is then plotted as scaled blocks on the line showing wind direction (Fig. 5.1).

Chapter 6
Measurement of Sunshine Duration

Abstract Sunshine duration is the length of time that the ground surface is irradiated by direct solar radiation (i.e., sunlight reaching the earth's surface directly from the sun). The instruments and accessories are required for measurement of sunshine duration have been explained. The installation of these instruments as well as the procedure for taking the observations has been discussed.

Keywords Sunshine · Measurement · Campbell-Stokes sunshine recorder Installation

6.1 Introduction

Sunshine duration is the length of time that the ground surface is irradiated by direct solar radiation (i.e., sunlight reaching the earth's surface directly from the sun). Sunshine duration is defined as "the period during which direct solar irradiance exceeds a threshold value of 120 watts per square meter (W/m^2)" (WMO 2003). This value is equivalent to the level of solar irradiance shortly after sunrise or shortly before sunset in cloud-free conditions.

6.2 Measurement of Sunshine Duration

For the measurement of sunshine duration, following instruments and accessories are required:

i. Campbell-Stokes sunshine recorder.
ii. Sunshine card.
iii. A special plastic scale.

© Springer International Publishing AG 2017
L. Ahmad et al., *Experimental Agrometeorology: A Practical Manual*,
https://doi.org/10.1007/978-3-319-69185-5_6

Fig. 6.1 Campbell Stokes sunshine recorder

Campbell-Stokes Sunshine Recorder

The Campbell-Stokes Sunshine Recorder consists of a 10 cm diameter glass sphere mounted on a spherical bowl. There are three partially overlapping grooves in the bowl in which three different types of cards are placed. The glass sphere is focused so that an image of the sun is formed on recording paper (Fig. 6.1). Three different recording cards are used depending on the season.

Recording Cards

The cards are made up of a good quality pasteboard of 0.04 mm thickness and are blue in color. This color enables the card to absorb the radiation and gives a good contrast when burned. There are three different types of cards (Fig. 6.2). The short curved are used for winter season, long curved card for summer season and straight cards in equinoxes are used.

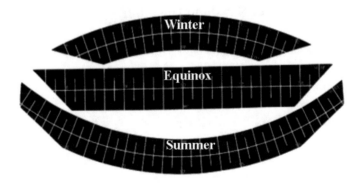

Fig. 6.2 Sunshine cards for different seasons

Cards	Season	Period	Grooves
Short curved	Winter	15th October to end of February	Upper
Long curved	Summer	12th April to 2nd September	Lower
Straight	Equinoxes	3rd September to 14th October; 1st March to 11th April	Middle

Special Plastic Scale

This is a type of time scale used to measure the length of burn to obtain the duration of sunshine. The scale is made up of celluloid and each hour is divided into 10 parts consisting of 0.1 h or 6 min. The parallel sunshine scale is used for straight cards and trapezoidal scale is used for long and short curved cards.

Installation and Working

The instrument is installed on a masonry pillar of 10 ft. (3.04 m) above the ground. The sphere is supported on the bowl according to the latitude of the place where the observatory is located.

The focus of the sphere shifts as the sun moves, and a burn trace is left on the recording card at the focal point. A burn trace at a particular point indicates the presence of sunshine at that time, and the recording card is scaled with hour marks so that the exact time of sunshine occurrence can be ascertained. Measuring the overall length of burn traces reveals the sunshine duration for that day. The total length of the burn is measured with the help of the special plastic scale.

Procedure

1. A suitable card depending on the season is set in the suitable groove.
2. Insert the card in the appropriate groove of the recorder such that the 12 h line coincides with the noon mark engraved on the bowl.
3. After sunset remove the burnt card.
4. Measure the burns using the special plastic scale.
5. Add up the values for all the hours and determine the total duration of sunshine hours for the day.

Chapter 7
Measurement of Solar Radiation

Abstract Electromagnetic radiation is emitted by everything in nature and the energy emitted by the sun is solar radiation. The chapter gives a description of solar radiation and different forms of solar radiation. Various instruments for measurement of solar radiation like, pyrheliometer, pyranometer and net radiometer have been discussed. The construction and working of quantum sensor, spectroradiometer, and infrared thermometer has also been described. Measurement of reflected radiation using pyrano-albedometer and tube solarimeter is given. Estimation of radiation intensity using BSS (angstrom formula) has also been included.

Keywords Solar radiation · Measurement · Units · Albedo estimation

7.1 Introduction

Electromagnetic radiation is emitted by everything in nature and the energy emitted by the sun is solar radiation. The sun emits a nearly constant amount of solar radiation (1.361 kW/m^2) continuously known as the solar constant. Solar constant is defined as "the amount of energy that normally falls on a unit area (1 m^2) of the earth's atmosphere per second when the earth is at its mean distance from the sun".

7.2 Forms of Solar Radiation

Direct Solar Radiation

The radiation received directly from the sun by the surface of the earth is direct solar radiation. It is also termed as incident radiation. Direct radiation has 45% Photo synthetically Active Radiation (PAR).

© Springer International Publishing AG 2017
L. Ahmad et al., *Experimental Agrometeorology: A Practical Manual*,
https://doi.org/10.1007/978-3-319-69185-5_7

Diffuse Radiation

The radiation scattered by the particles present in the atmosphere reaches the earth as diffuse radiation. The percentage of PAR in diffuse radiation is more than that of direct radiation being 65%.

Global Radiation

The total amount of solar radiation falling on a horizontal surface, i.e. the direct solar beam plus diffuse solar radiation on a horizontal surface is referred as global solar radiation.

Reflected Radiation

Reflected radiation is the part of radiation that occurs when radiation "bounces" off the target and is redirected. It is also known as albedo.

Thermal Radiation

Radiation of long wavelength that is emitted by the earth and heats the atmosphere is thermal or terrestrial radiation.

Net Radiation

It is the radiation balance between global and reflected solar radiation. In other words, it is the amount of energy available for processes at the ground surface.

7.3   Units

The solar irradiance is expressed in watts per square meter (W/m^2) and the total amount in joules per square meter (J/m^2). Formerly the unit of measurement was calorie. The conversion between calorie and joule is given as:

Solar irradiance	$1\ kW/m^2 = 1.433\ cal/cm^2/min$
Total amount of solar radiation	$1\ MJ/m^2 = 23.89\ cal/cm^2$

7.4   Measurement of Solar Radiation

Solar radiation is measured by different types of radiometers. A radiometer absorbs solar radiation at its sensor, transforms it into heat and measures the resulting amount of heat to ascertain the level of solar radiation. The radiometers used for ordinary observation are pyrheliometers and pyranometers that measure direct solar radiation and global solar radiation, respectively.

Pyrheliometer

The direct solar radiation is measured by means of a pyrheliometer. The receiving surface of a pyrheliometer is arranged normal to the solar direction. For this purpose, the instrument is generally mounted on a sun-tracking device called an equatorial mount.

Pyranometer

A pyranometer is used to measure global solar radiation falling on a horizontal surface. Its sensor has a horizontal radiation-sensing surface that absorbs solar radiation energy from the whole sky (i.e. a solid angle of 2π sr) and transforms this energy into heat. Global solar radiation can be ascertained by measuring this heat energy.

Net Radiometer

Net radiometer consists of two pyranometers whose sensors are exposed to earth and sky. The sky-facing sensor measures the incoming radiation while the sensor facing the earth measures outgoing radiation.

Quantum Sensor

Quantum sensor is used for the measurement of PAR. A tube solarimeter has several quantum sensors fixed at regular intervals on a tube. These sensors estimate the Intercepted PAR (IPAR) by the crop canopy. The quantum sensor measures the PAR received above the crop surface. The PAR reaching the soil surface is measured by means of tube solarimeters reaching the soil through the canopy. The difference between the IPAR and the PAR near the soil gives the intercepted PAR or PAR absorbed by the crop. It is measured in Einstein units (Ei) which are equal to one mole of protons.

Spectroradiometer

The spectroradiometer was developed by Indian Space Research Organization (ISRO), Bangalore, which measures radiation at an interval of 20 nm bandwidth between 400 and 1010 µm wavelength range. This instrument is used to find out the relationship of crop characters and reflectance.

Infrared Thermometer

The infrared thermometer senses the radiation in the IR region, i.e., radiation having a wavelength of 8–14 µm. it is used for the measurement of plant canopy temperature. The water status of plants can be estimated and accordingly irrigation can be scheduled.

7.5 Measurement of Incident and Reflected Solar Radiation

Albedo-Pyranometer

The albedo-pyranometer consists of two identical hemispherical glass domes each containing a sensor. One dome faces upwards and the other downwards. The upper dome (pyranometer) measures incident solar radiations whereas the lower dome (albedometer) measures the reflected short-wave solar radiation from the surface (Fig. 7.1). The sensor portion is made of nearly 100 copper-constantan thermocouples arranged circularly. These are imbedded on a substrate which contains Al_2O_3 which has high thermal conductivity. Differential heating of the thermocouple junction by irradiance generates emf, which is proportional to the incident energy. The emf is measured by the multi-voltmeter as the voltage output. The multi-voltmeter measures the output reading in millivolts (mV), which can be converted into Wm. The calibration factor for the instrument is given below:

1 mV = 0.083372 cal cm^{-2} min^{-1}
1 cal cm^{-2} min^{-1} = 697.674 Wm^{-2}
Irradiance (Wm) = Measured voltage (mV) × 58.16

The solar radiation in the wavelength range of 305–2800 nm is measured by the pyrano-albedometer. There are four connections of different colors, viz. green, yellow, white and brown. The green-yellow junctions obtain readings from the

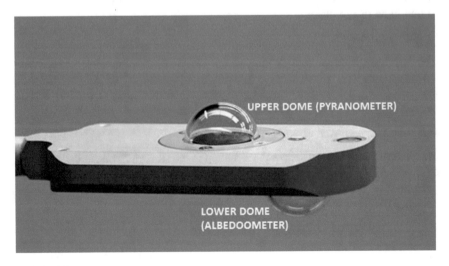

Fig. 7.1 Albedo-pyranometer

upper sensor (pyranometer) while the readings from the lower sensor (albedometer) are obtained with white-brown junctions. When the sensor is held horizontally, the total incident solar radiation or reflected short wave radiation between 305 and 2800 nm is measured as voltage output from above colored junctions with the help of a multi-voltmeter. When the sensor is placed in shade, then it gives the diffused solar radiation between 305 and 2800 nm.

Tube Solarimeter

Total solar radiation is measured by tube solarimeter also known as pyranometer. It has a tube shaped sensor composed of alternate black and white strips. The sensor consists of copper-constantan thermopile encased within a tube made from Pyrex Borosilicate Glass. This envelope limits the sensor response to visible and near infra-red radiation in the waveband 350–2500 nm (Fig. 7.2).

The incident energy flux results in a small temperature difference between the black and white areas. The copper-constantan thermopile converts this energy difference into output voltage. The black and white areas are alternated so that when radiation heats one side of the sensor more than the other, the mean temperature difference between black and white surfaces is not affected. The voltage output is measured with a multi-voltmeter (in millivolts). The calibration factor for tube solarimeter is 15 mV/kW m^{-2}. This can be denoted as:

$$Irradiance\,(\mathrm{kWm}^{-2}) = \frac{Measured\;Voltage\,(\mathrm{mV})}{15}$$

In order to measure the incident solar radiation, the tube solarimeter is held horizontally with its sensor face upwards. A spirit level on the sensor ensures its horizontal position. The total incident solar radiation is measured as the voltage output with the help of multi-voltmeter connected to it. Diffuse solar radiation can be measured when the sensor is placed in shade facing upwards. For measuring reflected solar radiation, the senor is inverted facing downwards.

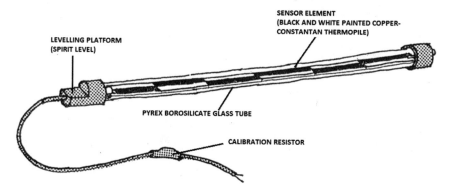

Fig. 7.2 Tube solarimeter

Estimation of Albedo from Total and Reflected Solar Radiation

'Albedo' is a Latin word which means whiteness. The albedo of a surface is the fraction of incident sunlight that it reflects. Albedo is a reflection coefficient and has a value less than one. Albedo is estimated from total solar radiation incident on a surface and radiation reflected by it. These radiations can be measured simultaneously by albedo-pyranometer or a tube solarimeter.

Albedo in percentage is calculated as:

$$Albedo\ (\%) = \frac{Reflected\ solar\ radiation\ in\ (Wm^2)}{Total\ incident\ solar\ radiation\ (Wm^2)} \times 100$$

Solar radiation transmitted within the crop canopy can be measured by placing the pyranometer/tube solarimeter facing upwards at the required depth in the canopy. Solar radiation interception (I) in percent at depth "x" is given as:

$$I = \frac{S - (A + T)}{S} \times 100$$

where,

S   Solar radiation incident at the top of canopy
A   Albedo
T   Transmission at depth "x".

Observations to be Recorded

The following readings should be taken in crop canopy:

1. Incident total solar radiation (Wm^{-2}).
2. Incident diffused solar radiation (Wm^{-2}).
3. Reflected solar radiation or albedo (%) from:

    (a) Bare soil
    (b) Grass surface
    (c) Metallic road
    (d) Water surface
    (e) Crop surface.

4. Transmitted solar radiation at different canopy depths (Wm^{-2}).
5. Solar radiation intercepted at different crop depths (%).

7.6 Measurement of Net Radiation

The difference between total upward and downward fluxes is net radiation. It is a measure of the energy available at the ground surface. The processes of evaporation, air and soil heating and energy consuming processes such as photosynthesis are also driven by Net Radiation. An expression for net radiation is given as:

$$R_n = (S\downarrow -S\uparrow) + (L\downarrow -L\uparrow)$$

$$\text{or}\quad R_n = S_n - L_n$$

where,

R_n net radiation.
S_n net shortwave radiation.
$S\downarrow$ incoming shortwave radiation.
$S\uparrow$ outgoing shortwave radiation.
L_n net longwave radiation.
$L\downarrow$ incoming longwave radiation.
$L\uparrow$ outgoing longwave radiation.

Net radiation is measured by a net pyrradiometer (Fig. 7.3). A net radiometer has two plastic domes which allow radiation directed towards and away from the earth to be transmitted. The domes are semi-rigid and are kept inflated by blowing air into them through a check valve in the handle. The handle contains silica gel which acts as a desiccant.

The sensor of the net pyrradiometer is a differential thermopile which uses use temperature differences between black body surfaces (top and bottom) to determine the difference between incoming solar radiation and reflected radiation from the earth's surface. The difference in temperature is converted to an output voltage difference. Diagonal white lines, which are transparent to long wave radiation, are painted on the plates. These lines reflect short wave radiations and balance the spectral response of the instrument.

The instrument is held horizontally at a height of 50–100 cm over the measurement surface. The horizontal position is ensured by a circular level on the instrument. A multi-voltmeter records the voltage once the domes are fully inflated. Voltage is converted into energy as:

1 mV = 0.083372 cal cm^{-2} min^{-1}
1 cal cm^{-2} min^{-1} = 697.674 Wm^{-2}

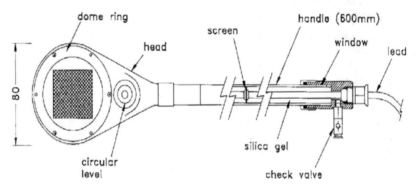

Fig. 7.3 Net pyrradiometer

Observations to Be Recorded

The following readings should be taken at different surfaces:

1. R_n at grass surface.
2. R_n at bare soil surface.
3. R_n at crop canopy surface.

7.7   Measurement of Photosynthetically Active Radiation

The portion of the electromagnetic spectrum used by the plants for the process of photosynthesis is known as Photo synthetically Active Radiation. The wavelength lies in the wave band of 400–700 nm. PAR can be measured in energy units (watts m^{-2}). PAR is measured by means of Line Quantum Sensor.

Line Quantum Sensor

Line Quantum Sensor is usually a rod of 1 m length and 12.7 mm in width. A number of high stability silicone photo voltaic detectors are arranged linearly along its length at a distance of 2.38 cm from each other (Fig. 7.4). A water proof anodized aluminium case encloses the detectors. The case has an acrylic diffuser and stainless hardware.

The output from the sensor is connected to an integrating quantum meter along with a calibrated connector (provided with the sensor). Both instantaneous values as well as the values integrated over 10, 100 and 1000 s can be read from the meter. The effect of rapidly changing cloud cover, surface waves, movement of sensor etc. is removed by using average values. The values are displayed on an LCD display. A continuous recorder can also be connected to the reader. The output of a quantum sensor is given in $\mu E\ s^{-1}\ m^{-2}$. For natural daylight conditions,

$1\ Wm^{-2} = 0.2174 \times \mu E\ s^{-1}\ m^{-2}$	
Also, $1\ \mu Es^{-1}\ m^{-2} = 1\ \mu mol\ s^{-1}\ m^{-2}$	
$= 6.02 \times 10^{17}\ photons\ s^{-1}\ m^{-2}$	
$= 6.02 \times 10^{17}\ quanta\ s^{-1}\ m^{-2}$	

In order to measure the IPAR (Intercepted), the line quantum sensor is placed above the crop canopy or in the open. The sensor is inverted over the crop canopy to get RPAR (Reflected). For measuring TPAR, the sensor is placed perpendicular to the row direction of the crop horizontally at the ground level. APR is then estimated as:

$$APAR = IPAR - (RPAR + TPAR)$$

The values are expressed in terms of percent of IPAR. For the measurement of profile APAR, the sensor is placed at 25, 50 and 75% of the crop height and TPAR is recorded at the respective heights.

It is generally advisable to take readings between 1100 and 1300 h IST for one reading per day. In case of diurnal observation requirements, hourly readings should be noted from sunrise to sunset.

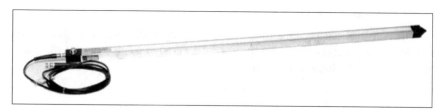

Fig. 7.4 Line quantum sensor

Observations

The following readings are to be taken in a crop canopy:

1. IPAR.
2. RPAR.
3. TPAR.

Readings taken to measure profile APAR:

1. TPAR at 25% crop height.
2. TPAR at 50% crop height.
3. TPAR at 75% crop height.

7.8   Estimation of Radiation Intensity Using BSS (Angstrom Formula)

Various climate models have been developed for use in predicting the monthly average global solar radiation, the popular one being the Angstrom-type regression equation developed by Angstrom in 1924. This relates monthly average daily global radiation to the average daily sunshine hours, and is given by the following expression:

$$\frac{Q}{Q_o} = a + b \frac{N}{N_o}$$

where,

Q   Monthly average daily global radiation on a horizontal surface (MJ m^{-2} day^{-1}),

Q_o Monthly average daily extraterrestrial radiation on a horizontal surface (MJ m^{-2} day^{-1}) (Appendix B),

a, b   Constants, whose value depends on location and month of observation,

N   Monthly average daily number of hours of bright sunshine (BSS) (Appendix B),

N_o Monthly average daily maximum number of hours of possible sunshine (or day length) given in Table…

Although it is not possible to estimate the daily total amount of global solar radiation on a particular day from the sunshine duration using this method, it does enable rough estimation of a monthly value.

Chapter 8
Measurement of Cloud Cover

Abstract Cloud is an aggregate of very small water droplets, ice crystals, or a mixture of both, with its base above the Earth's surface, which is perceivable from the observation location. The limiting liquid particle diameter is of the order of 200 μm; drops larger than this comprise drizzle or rain. The chapter describes types of clouds on the basis of their height and appearance. The estimation of cloud cover for meteorological purposes has been given.

Keywords Cloud · Measurement · Types

8.1 Introduction

Cloud is an aggregate of very small water droplets, ice crystals, or a mixture of both, with its base above the Earth's surface, which is perceivable from the observation location. The limiting liquid particle diameter is of the order of 200 μm; drops larger than this comprise drizzle or rain.

8.2 Types of Clouds

Clouds are classified according to their height above and appearance (texture) from the ground.

On the basis of formation at different heights in the sky above surface these may be high clouds (C_H), medium clouds (C_M) or low clouds (C_L).

On the basis of appearance, the clouds are classified as:

1. Cirro: curl of hair;
2. Alto: mid;
3. Strato: layer;
4. Nimbo: rain, precipitation; and
5. Cumulo: heap.

© Springer International Publishing AG 2017
L. Ahmad et al., *Experimental Agrometeorology: A Practical Manual*,
https://doi.org/10.1007/978-3-319-69185-5_8

| CIRRUS | CIRROSTRATUS | CIRROCUMULUS |

Fig. 8.1 High-level clouds

High Clouds

High-level clouds occur above about 20,000 ft and are given the prefix "cirro". Due to cold tropospheric temperatures at these levels, the clouds primarily are composed of ice crystals, and often appear thin, streaky, and white (although a low sun angle, e.g., near sunset, can create an array of color on the clouds). The three main types of high clouds are cirrus, cirrostratus, and cirrocumulus (Fig. 8.1).

Medium Clouds

The bases of clouds in the middle level of the troposphere, given the prefix "alto," appear between 6500 and 20,000 ft. Depending on the altitude, time of year, and vertical temperature structure of the troposphere, these clouds may be composed of liquid water droplets, ice crystals, or a combination of the two, including super-cooled droplets (i.e., liquid droplets whose temperatures are below freezing). The two main type of mid-level clouds are altostratus and altocumulus (Fig. 8.2).

Low Clouds

Low-level clouds are not given a prefix, although their names are derived from "strato" or "cumulo", depending on their characteristics. Low clouds occur below 6500 ft, and normally consist of liquid water droplets or even super-cooled droplets,

| ALTOSTRATUS | ALTOCUMULUS |

Fig. 8.2 Medium clouds

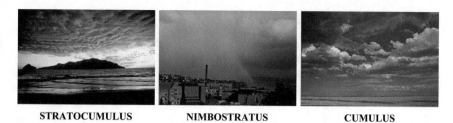

STRATOCUMULUS   **NIMBOSTRATUS**   **CUMULUS**

Fig. 8.3 Low-level clouds

except during cold winter storms when ice crystals (and snow) comprise much of
the clouds. The two main types of low clouds include stratus, which develop
horizontally, and cumulus, which develop vertically (Fig. 8.3).

Stratocumulus clouds are hybrids of layered stratus and cellular cumulus.
Stratocumulus also can be thought of as a layer of cloud clumps with thick and thin
areas. Thick, dense stratus or stratocumulus clouds producing steady rain or snow
often are referred to as nimbostratus clouds. Cumulus clouds are more cellular
(individual) in nature, have flat bottoms and rounded tops, and grow vertically.

Fig. 8.4 Scale of cloud cover
measurement. *Note* 0 oktas
represents the complete
absence of cloud. 1 okta
represents a cloud amount of
1 eighth or less, but not zero.
7 oktas represents a cloud
amount of 7 eighths or more,
but not full cloud cover.
8 oktas represents full cloud
cover with no breaks. 9 oktas
represents sky obscured by
fog or other meteorological
phenomena

Cloud Cover

Symbol   Scale in oktas (eighths)

Symbol	Scale	Description
○	0	Sky completely clear
⊕	1	
◑	2	
◕	3	
◐	4	Sky half cloudy
⊖	5	
◕	6	
◑	7	
●	8	Sky completely cloudy
⊗	(9)	Sky obstructed from view

8.3   Estimation of Cloud Cover

There is no instrument for measuring the cloud amount in the sky. Cloud amount is expressed in terms of "Okta" or one-eighth of the sky. The assessment of the total amount of clouds therefore consists of estimating how much total area of the sky nearest to the earth is covered with clouds. The observer assumes the sky to be divided into four quadrants. The clouds scattered in different parts of the sky are supposed to be present together. The observer then estimates the number of quadrants the clouds occupy. Cloud amount is reported in oktas. Scale of cloud cover measured in oktas (eighths) with the meteorological symbol for each okta is shown in Fig. 8.4.

Chapter 9
Measurement of Precipitation

Abstract Precipitation denotes all forms of water (liquid or solid) that reach the earth from atmosphere. Precipitation includes rain, snow, hail, dew, fog, drizzle etc. Of all these only rain and snow contribute significant amount of water on the earth. The chapter explains different forms of precipitation and the instruments and techniques associated with their measurement. Different methods used for analysis, presentation and trend analysis of rainfall data are discussed.

Keywords Rainfall · Snow · Measurement · Units · Types

9.1 Introduction

The term precipitation denotes all forms of water (liquid or solid) that reach the earth from atmosphere. Precipitation includes rain, snow, hail, dew, fog, drizzle etc. Of all these only rain and snow contribute significant amount of water on the earth. The essential requirements for the precipitation to occur are:

  i. The atmosphere must have moisture,
 ii. Lifting of air mass in the atmosphere, i.e. cooling and condensation of moisture,
iii. The atmosphere must have sufficient condensation nuclei to aid condensation.

9.2 Forms of Precipitation

Different combinations of states of water, temperature and wind conditions give rise to the various forms of precipitation.

  i. **Drizzle**: A fine sprinkle of very small and rather uniform water droplets with very low intensity.
 ii. **Rain**: Rain is precipitation in the form of liquid water drops.

© Springer International Publishing AG 2017   55
L. Ahmad et al., *Experimental Agrometeorology: A Practical Manual*,
https://doi.org/10.1007/978-3-319-69185-5_9

iii. **Snow**: Snow is precipitation comprising ice crystals which are either translucent or white. For meteorological records, snow is melted and its amount is expressed in terms of the equivalent depth of rain.
 iv. **Sleet**: Sleet is precipitation of melting snow or a mixture of snow and rain.
  v. **Glaze**: The ice coating formed when rain or drizzle freezes as it comes in contact with cold objects at the ground is called glaze.
 vi. **Hail**: Hail is precipitation of irregular balls or lumps of ice which fall from cumulonimbus clouds and are often associated with thunderstorms.
vii. **Dew**: Dew is formed at late nights or in early mornings. It appears as beautiful globules of clear water on grass blades, flower petals and other objects on the ground.

9.3 Terminology and Units

Precipitation is expressed in terms of the depth to which water would stand on an area if all the rain (precipitation) were collected in it. The magnitude of precipitation is expressed in terms of millimeters. Various terms associated with the measurement and analysis of precipitation are given in Table 9.1.

9.4 Measurement of Rainfall

Rainfall is the principle form of precipitation in India. Rainfall is collected and measured by means of rain gauges. The rainfall measurement is taken as the vertical depth of water that would accumulate on a level surface if the rainfall remains where it fell. Rain gauges are primarily of two types:

 i. Ordinary or non-recording rain gauge.
ii. Recording Rain Gauge.

Table 9.1 Description and units of some common terms associated with measurement and analysis of rainfall

Term	Description	Units
Depth	Depth to which water would stand on an area if all the rain (precipitation) were collected in it	mm, cm, inches
Intensity	The quantity of rain falling in a given time	m/h, cm/h, mm/h
Duration	The period of time during which rain falls	h, minutes
Frequency	The expectation that a given depth of rainfall will fall in a given time	once in three years, once in five years, etc.

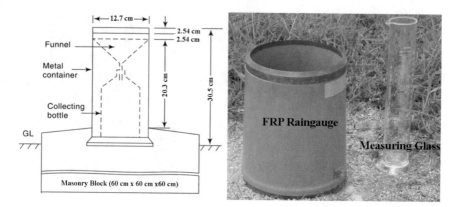

Fig. 9.1 Symon's gauge or ordinary rain gauge

Ordinary Rain Gauge

The non-recording rain gauge used in India is the Symon's gauge. The rain gauge has a circular collecting area of 12.7 cm (5.0 in.) diameter connected to a funnel. The collector rim is set in a horizontal plane at a height of 30.5 cm above the ground level (Fig. 9.1). The base and collector are made of fiber reinforced plastic (FRP), hence this rain gauge is also known as FRP rain gauge. The rainfall catch is discharged into a receiving vessel by means of the funnel. The funnel as well as the receiving vessel are housed in a metallic container. Water collected in the receiving vessel is measured by a graduated cylinder having an accuracy up to 0.1 mm.

For uniformity, the rainfall is measured every day at 8.30 AM (IST) and is recorded as the rainfall of that day. The receiving bottle normally does not hold more than 10 cm of rain and as such in the case of heavy rainfall the measurements must be done more frequently and entered. However, the last reading must be taken at 8.30 AM and the sum of the previous readings in the past 24 h entered as total of that day. Proper care, maintenance and inspection of rain gauges, especially during dry season.

Ordinary rain gauge can also be used to measure snowfall as discussed in Sect. 9.5.

Recording Rain Gauge

Recording gauges give a continuous plot of rainfall against time and provide valuable data of intensity and duration of rainfall. Some of the commonly used rain gauges are:

a. Tipping Bucket Type

This rain gauge has been adopted for use by USWB. It is a 30 cm rain gauge where the rain from the funnel falls on to a pair of small buckets. The buckets are balanced in such a manner that when 0.25 mm rainfall collects in one bucket, it tips and brings another in position (Fig. 9.2). The water tipped from the buckets is stored in

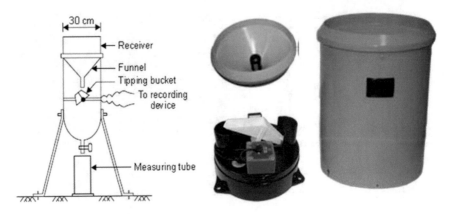

Fig. 9.2 Tipping bucket rain gauge

a storage can. The tipping starts an electrically driven pen which traces the record on clockwork-driven chart. Total rainfall is determined by measuring the water collected in the storage can.

b. Weighing Bucket Type

In this rain gauge, the catch from the funnel empties into a bucket mounted on a weighing scale. A clock-work driven chart records the weight of the bucket and its contents (Fig. 9.3). This rain gauge gives the plot of accumulated rainfall versus time. In some instruments of this type the recording unit is so constructed that the pen reverses its direction at every preset value (20 cm in Fig. 9.4) in order to obtain a continuous plot of storm.

c. Natural Siphon Type

Also known as float type rain gauge, this rain gauge has a float chamber into which the catch from the funnel drains. The rainfall collected causes the float to rise which

Fig. 9.3 Weighing bucket rain gauge

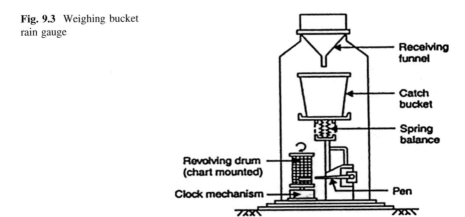

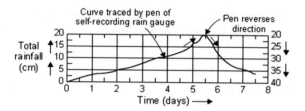

Fig. 9.4 Plot of a weighing bucket rain gauge

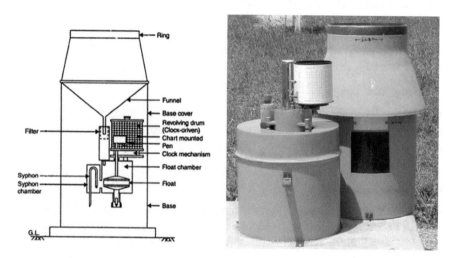

Fig. 9.5   Natural siphon type rain gauge

initiates a pen to trace the record on a rotating drum driven by a clock-work mechanism. When the pre-set maximum level is reached by the float, a siphon mechanism empties the float chamber (Fig. 9.5). In India, this rain gauge is adopted as the standard recording-type rain gauge.

A typical chart from this type of rain gauge is shown in Fig. 9.6. The vertical lines in the pen-trace correspond to the sudden emptying of the float chamber by

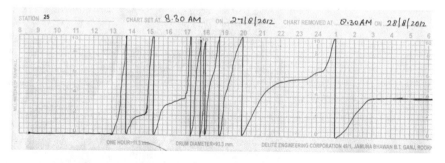

Fig. 9.6   Plot of a weighing bucket rain gauge

syphon action which resets the pen to zero level. The natural syphon-type recording rain gauge gives a plot of the mass curve of rainfall.

9.5 Measurement of Snow

Snow gauges are used for the measurement of snowfall and the water equivalent of snow is taken as the depth of precipitation. Snowfall differs from rainfall in that it may accumulate on the surface before it melts and causes runoff. The magnitude of snowfall is expressed in terms of "water equivalent of snow". Water equivalent of snow is the depth of water that would result in melting a unit of snow. Water equivalent of snow is obtained by any of the following ways:

Ordinary Rain Gauge

Ordinary rain gauge can be used to measure snowfall also. When snow is expected, the funnel and receiving bottle are removed and the snow is allowed to collect in the outer metal container. The snow is then melted and the depth of resulting water measured.

Snow Boards

Snowboards are light colored, square, flat plywood pieces having a side of about 41–61 cm and a thickness of about 1.3–1.9 cm. snow samples are cut off from the board to obtain the water equivalent of snow (Fig. 9.7).

Snow Gauges

Snow gauges have a large (203 mm) cylindrical receiver which collects the snow as it falls. The receiver is usually mounted on a tower to keep its rim above the anticipated maximum depth of accumulated snow (Fig. 9.8). The snow collected is

Fig. 9.7 Snow board

Fig. 9.8 Snow gauge

then melted by adding a premeasured quantity of hot water to obtain the water equivalent of snow.

9.6 Measurement of Hail

Hail is usually measured by means of rain gauge. The outer cylinder gathers hail, while melted water is collected in the inner cylinder/collecting vessel. The intensity of the hail storm is measured by means of hail pads. A hail pad consists of a 12″ by 12″ square of Styrofoam covered in Heavy Duty Aluminum foil. When the hail hits the pad it leaves an indentation and from this the specialized meteorologists can calculate the size, force, direction, and intensity of the hail storm.

9.7 Measurement of Dew

Dew is measured by means of Duvdevani Dew gauge and dew album. Measurement of dew is important because absorption of dew is an important factor in survival of natural vegetation in arid regions. Absorption of dew by leaves is beneficial for crops and reduces transpiration losses.

The Duvdevani dew gauge is a specially treated wooden bar ($32 \times 5 \times 2.5$ cm). The bar is coated with red oxide. The bars are exposed on a stand at heights of 5, 10 and 15 cm just before sunset (Fig. 9.9). The amount of dew is generally expressed in g/100 cm^2 or in mm of dew. One mm of dew is equal to 10 g of dew per 100 cm^2.

Fig. 9.9 Duvdevani dew gauge

The size of the dew droplets formed on the bar is visually compared to photographs of known dew quantities before sunrise. First the dew deposited on the upper surface of the bar is noted, then the bar is inverted and dew accumulation on the lower surface are also observed. The mean of lower and upper surface deposits for each block is recorded. Water equivalent of dew is then estimated from dew scale (Table 9.2).

Table 9.2 Table for estimation of water equivalent of dew

Dew scale number	Water equivalent (mm)
0	No dew
1	0.020
2a, 2b	0.045
3a, 3b	0.075
4a, 4b	0.110
5a, 5b	0.160
6a, 6b	0.210
7a, 7b	0.270
8	0.350
9	No dew but rain

9.8 Analysis of Rainfall Data

Before using rainfall records at a station, the data has to be checked for its continuity. Gaps may be present in data due to instrument failure, absence of observer etc. Filling of missing data involves transmittal of rainfall amounts observed at nearby index station to station with missing station. Commonly used methods are:

- i. Arithmetic Average Method.
- ii. Normal Ratio Method.
- iii. Inverse Distance Method.

i. **Arithmetic Average Method**

This method is used when the normal annual rainfall of missing station is within 10% of surrounding stations, data of at least 3 surrounding stations are available and stations are evenly spaced.

$$P_x = \frac{1}{m}(P_1 + P_2 + \cdots + P_m)$$

where,

P_x Estimated rainfall at missing station,

P_1, P_2, P_m Observed rainfall at surrounding 'm' index stations.

ii. **Normal Ratio Method**

This method is used if normal annual precipitation of stations differs more than 10% of the missing station. Usually 3 or more surrounding stations are selected. Missing data is estimated as:

$$P_x = \frac{N_x}{m}\left(\frac{P_1}{N_1} + \frac{P_2}{N_2} + \cdots + \frac{P_m}{N_3}\right)$$

where,

N_x Normal Annual Rainfall of missing station,

N_1, N_2, N_m Normal Annual Rainfall at surrounding 'm' index stations.

iii. **Inverse Distance Method**

In this case, weight of surrounding stations on the basis of their distance from gauge with missing data is considered. The distance is computed by establishing a set of axis running through gauge (station A) with missing data (Fig. 9.10).

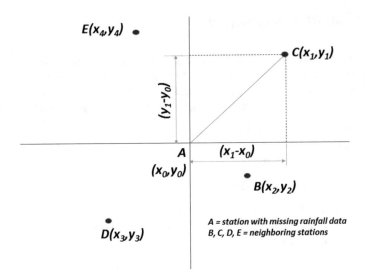

Fig. 9.10 Inverse distance method

The squared distance between origin and station 'i' is computed as:

$$D_i^2 = (x_i - x_0)^2 + (y_i - y_0)^2$$

The rainfall at station A is calculated as:

$$P_A = \frac{\sum_{i=1}^{n} P_i \times W_i}{\sum_{i=1}^{n} W_i}$$

where,

$$W_i = \frac{1}{D_i^2}$$

9.9   Presentation of Rainfall Data

A few methods generally used for presentation of rainfall data are given below. These methods have been found useful for interpretation and analysis of rainfall data.

1. Mass curve of rainfall.
2. Hyetograph.

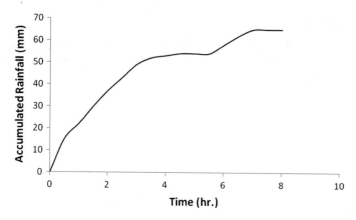

Fig. 9.11 Mass curve of rainfall

i. **Mass curve of rainfall**

It is a plot of accumulated precipitation against time in chronological order (Fig. 9.11). This gives information about duration and magnitude of a rainfall event and can be used to obtain the rainfall intensity at different time periods.

Fig. 9.12 Hyetograph

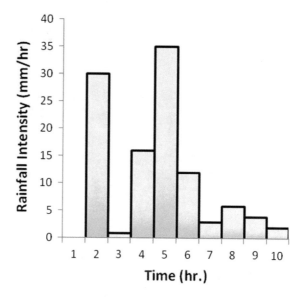

ii. **Hyetograph**

It is a plot of rainfall intensity against time interval. It is represented by means of a bar chart (Fig. 9.12). The area under the hyetograph represents the total rainfall depth in that period.

Chapter 10
Measurement of Evaporation

Abstract Evaporation is an important component of hydrological cycle. It is a major factor which determines the irrigation water requirements of crops. Hence, the observation on evaporation is important for studying crop growth. The chapter gives a description of terms and units associated with evaporation and evapotranspiration. Various instruments and techniques used for measurement of evaporation and evapotranspiration have been discussed. Various empirical methods which estimate evapotranspiration on the basis of meteorological parameters are also included in the chapter.

Keywords Evaporation · Measurement · Estimation · Units

10.1 Introduction

The process of change of a liquid to gaseous state at a free surface below its boiling point is known as evaporation. It is an important component of hydrological cycle. It is a major factor which determines the irrigation water requirements of crops. Hence, the observation on evaporation is important for studying crop growth.

10.2 Terminology

The terms associated with evaporation are defined below. These definitions have been presented by International Glossary of Hydrology (WMO/UNESCO 1992) and the International Meteorological Vocabulary (WMO 1992).

   i. **Actual Evaporation**: Quantity of water evaporated from an open water surface or from the ground.

  ii. **Transpiration**: Process by which water from vegetation is transferred into the atmosphere in the form of vapor.

© Springer International Publishing AG 2017
L. Ahmad et al., *Experimental Agrometeorology: A Practical Manual*,
https://doi.org/10.1007/978-3-319-69185-5_10

iii. **Actual evapotranspiration (or effective evapotranspiration)**: Quantity of water vapor evaporated from the soil and plants when the ground is at its natural moisture content.

iv. **Potential evaporation (or evaporativity)**: Quantity of water vapor which could be emitted by a surface of pure water, per unit surface area and unit time, under existing atmospheric conditions.

v. **Potential evapotranspiration**: Maximum quantity of water capable of being evaporated in a given climate from a continuous expanse of vegetation covering the whole ground and well supplied with water. It includes evaporation from the soil and transpiration from the vegetation from a specific region in a specific time interval, expressed as depth of water.

vi. **Reference evapotranspiration**: It is the evapotranspiration from a reference surface. The reference surface is a hypothetical grass reference crop with an assumed crop height of O. 12 m, a defined fixed surface resistance of 70 s m^{-1} and an albedo of 0.23. The reference surface closely resembles an extensive surface of green, well-watered grass of uniform height, actively growing and completely shading the ground.

10.3   Units

The rate of evaporation is defined as the amount of water evaporated from a unit surface area per unit of time. It can be expressed as the mass or volume of liquid water evaporated per area in unit of time, usually as the equivalent depth of liquid water evaporated per unit of time from the whole area. The unit of time is normally a day. The amount of evaporation should be read in millimeters (WMO 2003).

10.4   Measurement of Evaporation

Evaporation is measured by means of evaporimeters. Evaporimeters are pans containing water which are exposed to atmosphere. The loss of water by evaporation is measured at regular intervals. The most common pan evaporimeters used in India for the measurement of evaporation is USWB (United States Weather Bureau) Class a Land Pan (Class A pan evaporimeter).

The Class a land pan is a standard pan with a diameter of 1210 mm and a depth of 255 mm. The depth of water is maintained between 18 and 20 cm (Fig. 10.1). The pan is normally made of unpainted galvanized iron sheet painted white. The pan is placed on a wooden platform of 15 cm height above the ground to allow free circulation of air below the pan. The pan is covered by a wire mesh to avoid birds from drinking water from the pan. Evaporation measurements are made by

Fig. 10.1 USWB class A land pan

measuring the depth of water with a hook gauge in a stilling well or by a fixed-point gauge using measuring bucket.

The hook gauge consists of a movable scale fixed with a hook. The rotating head of the hook gauge is divided into 100 divisions so that the level of water may be read correct to 1/100 of an inch. The point of the hook indicates water level. The hook gauge rests on a stilling well of size 10 cm diameter and 30 cm height on which, is placed within the tank. It isolates some amount of the water surface in the pan so that it is not disturbed by waves produced by wind.

In fixed point gauge, a brass pointer is fixed vertically at the center of the cylindrical stilling well. The tip of the rod is located at 6–7 cm below the rim of the pan. There are three small holes at the bottom of the stilling well to permit the flow of water into and out of the well. For taking the reading, the water level is brought to the same fixed point by adding water using a graduated measuring bucket.

The ratio of the area of the bucket to the area of evaporation pan is exactly 1:100. It has is graduated from 0 to 20 along its height. The graduation runs from top to bottom in ascending order. One full cylinder of water rises 2 mm height in the pan. The evaporation can be measured correct to 0.1 mm.

10.5 Observations to be Recorded

Evaporation is measured every day at 8:30 am IST. The observations to be recorded are:

1. Temperature of water in the pan.
2. Amount of water added or removed to bring back level to the tip of the pointer.
3. Amount of rainfall during the past 24 h.

10.6 Measurement of Evaporation

Evaporation measurement is done for in different conditions.

No Rainfall

When there is no rainfall, measured quantity of water is added to the evaporation pan up to the tip of the pointer using the measuring bucket. The evaporation is equal to the amount of water added to the pan.

Rainfall with Pointer Above the Water Level

When there is rainfall, but the tip of the fixed point gauge remains above the water level, measured quantity of water is added by measuring bucket till the water level reaches tip of the pointer in the stilling well. Evaporation is estimated as:

$$Evaporation = Water\ added + Rainfall\ during\ the\ period$$

Rainfall with Pointer Below the Water Level

When rainfall is more than evaporation and the water level inside the stilling well is above the tip of the pointer, in this case water is removed from reservoir with the help of measuring bucket until the level of water comes to the tip of the pointer. Evaporation is estimated as:

$$Evaporation = Rainfall\ during\ the\ period - Water\ removed$$

Overflow

When there is heavy rainfall, over flow will take place from the pan, evaporation cannot be recorded.

10.7   Measurement of Evapotranspiration

For determination of evapotranspiration, lysimeters are used. A lysimeter is a special water tight tank containing a block of soil and set in a field of growing plants. The plants grown in the lysimeter are the same as in the field. Evapotranspiration is estimated in terms of the amount of water required to maintain constant moisture conditions within the tank measured either volumetrically or gravimetrically through an arrangement made in the lysimeter. Lysimeters should be designed to accurately reproduce the soil conditions, moisture content, type and size of the vegetation of the surrounding area. They should be so buried that the soil is at the same level inside and outside the container.

There are three types of lysimeters:

  i. Non-weighing type lysimeters.
 ii. Floating type lysimeters.
iii. Weighing type lysimeters.

Non-weighing Type Lysimeters

Non-weighing (percolation-type) lysimeters can be used only for measuring volumetric changes in water balance weekly or biweekly. They do not provide accurate daily estimates. A simple non-weighing lysimeter is shown in Fig. 10.2. Irrigation water is applied to the lysimeter. A layer of pebbles is placed at the bottom to facilitate easy drainage. Excess water is collected from below at a suitable distance.

Fig. 10.2 Non-weighing lysimeter

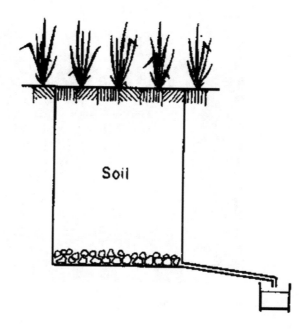

Soil

Weighing Type Lysimeters

Weighing lysimeters provide precise information on soil moisture changes for daily or even hourly periods. This lysimeter is mainly composed of an inner tank, which is placed inside another tank. The outer tank is in contact with the surrounding soil (Fig. 10.3). The inside container can be isolated from the outer tank and is weighed by scales. A heavy liquid like $ZnCl_2$ can be used to float the tank in water and the water gain or loss is estimated in terms of water displaced. This lysimeter is most commonly used for ET measurement but it has some important limitations. The restricted root growth, the disturbed soil structure in the lysimeter causing changes in water movement and possibly the tank temperature regime, resulting in condensation of water on the walls of the container. Harrold and Dreilbelbis (1967) estimated that errors due to dew formation were in the order of 250 mm per annum. Other limitations include the 'bouquet effect' whereby the canopy of the plants grown in the lysimeter is above and extends over the surrounding crop, resulting in a higher evapotranspiration rate. In spite of these limitations, it is the best technique for precise studies on evapotranspiration.

Floating Type Lysimeters

In this lysimeter, soil is placed in a hollow-walled tank that floats in water inside a slightly larger reservoir tank (Fig. 10.4). A loss or gain of weight in the floating tank results in a change of water level in the reservoir tank and stilling well. This level is recorded by a sensitive water-level recorder. Any overflow from an excessive weight on the floating tank is retained in an overflow tank at the recorder.

Fig. 10.3 Weighing lysimeter

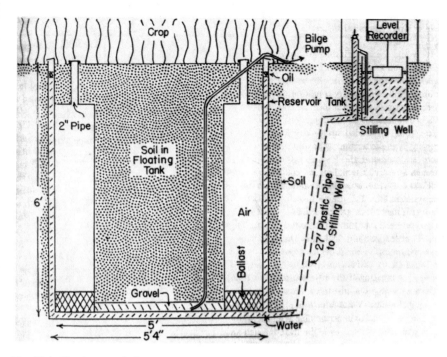

Fig. 10.4 Floating type lysimeters

Table 10.1 Some common methods of estimating reference ET

S.no.	Method of ET_0 estimation	Equations	Required meteorological data
1	FAO-24 Corrected Penman	$ET_0 = c\left[\frac{\Delta}{\Delta+\gamma}(R_n - G) + \frac{\gamma}{\Delta+\gamma}2.7W_f(e_s - e_a)\right]$	Net radiation, vapour pressure deficit and wind velocity
2	Priestley-Taylor (P-T)	$ET_0 = \propto \left[\frac{\Delta}{\Delta+\gamma}(R_n - G)\right]$	Net radiation, soil heat flux and vapour pressure deficit
3	FAO-24 Blaney-Criddle (F B-C)	$ET_0 = a + b[p \times 0.46\bar{T} + 8.13]$	Annual day time hours, temperature and wind velocity
4	FAO Pan Evaporation (F E-Pan)	$ET_0 = K_{pan} \times ET_{pan}$	Pan evaporation

(continued)

Table 10.1 (continued)

S.no.	Method of ET_0 estimation	Equations	Required meteorological data
5	Penman Monteith (P-Mon)	$$ET_0 = \frac{0.408\Delta(R_n - G) + \gamma\frac{900}{T_{mean}+273}U_2(e_s - e_a)}{\Delta + \gamma(1 + 0.34U_2)}$$	Vapour pressure deficit, radiation flux, wind velocity, temperature and soil heat flux

where,

ET_0 = references evapotranspiration [mm day^{-1}],

R_n = net radiation at the crop surface [MJ m^{-2} day^{-1}],

G = soil heat flux density [MJ m^{-2} day^{-1}],

T_{mean} = mean daily air temperature at 2 m height [°c],

U_2 = wind speed at 2 m height [m s^{-1}],

e_s = saturation vapor pressure [kpa],

e_a = actual vapor pressure [kpa],

$e_s - e_a$ = saturation vapor pressure deficit [kpa],

Δ = slop vapor pressure curve [kpa c^{-1}],

γ = psychrometric constant [kpa c^{-1}],

E_{pan} = pan evaporation [mm day^{-1}],

k_p = pan coefficient

10.8   Estimation of Evapotranspiration

The potential evaporation of any crop can be calculated by multiplying the reference crop evapotranspiration by a coefficient "K". The value of K depends upon the type of the crop and stage of growth. Some common methods of estimating Reference ET using meteorological data are tabulated in Table 10.1.

Chapter 11
Measurement of Atmospheric Pressure

Abstract The atmospheric pressure on a given surface is the force exerted per unit area by the weight of the Earth's atmosphere above. Atmospheric pressure is an important parameter for studying weather of a location and its interaction with the crops. The chapter includes various types of barometers used for the measurement of atmospheric pressure. The relation between atmospheric pressure and weather has been discussed in brief.

Keywords Atmospheric pressure · Units · Measurement · Barometer

11.1 Introduction

The atmospheric pressure on a given surface is the force exerted per unit area by the weight of the Earth's atmosphere above. Thus, the atmospheric pressure is equal to the weight of a vertical column of air above the Earth's surface, extending to the outer limits of the atmosphere.

11.2 Units

The basic unit for atmospheric pressure measurements is the Pascal (Pa) (or newton per square meter). Meteorological pressure measurements are reported in terms of hectopascals (hPa). 1 hPa is equal to 100 Pa. Also, 1 hPa is equal to 1 mb (millibar) that was used formerly.

Some barometers are graduated in the unit in Hg or mmHg. Under standard conditions, the pressure exerted by a pure mercury column which is 760 mm high is 1013.250 hPa, so the conversion factors are represented as follows:

© Springer International Publishing AG 2017
L. Ahmad et al., *Experimental Agrometeorology: A Practical Manual*,
https://doi.org/10.1007/978-3-319-69185-5_11

1 h Pa = 0.750062 mm Hg; = 0.029530 in Hg	
1 mm Hg = 1.333224 h Pa; = 0.03937008 in Hg	
1 in Hg = 33.8639 h Pa	
1 bar = 10^5 Pa	

11.3   Measurement of Atmospheric Pressure

Atmospheric pressure is measured by means of an instrument known as barometer. Some common types of barometers are described in the following sections.

Principle of Mercury Barometer

When a 1 m long, open ended glass tube is filled with mercury and is then turned upside down into a container filled with mercury, part of the mercury flows out of the glass tube into the container. "Torricelli an vacuum" is then produced at the top of the glass tube and the mercury level stabilizes at approximately 76 cm from the mercury level in the container (Fig. 11.1). Torricelli's experiment revealed that

Fig. 11.1 Torricelli's experiment

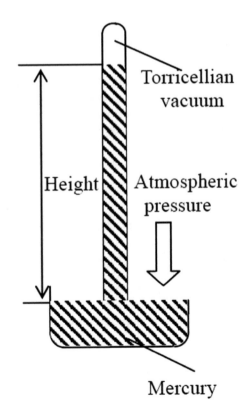

such a height indicates the ambient atmospheric pressure. The principle of mercury barometer is to measure atmospheric pressure from precise measurement of this height.

Fortin Barometer

It is a standard instrument for measurement of pressure consisting of three main parts: the mercury cistern (right), the glass barometer tube (center) and the scale (left) as shown in Fig. 11.2. The bottom of the mercury cistern is made of a wash-leather bag (sheepskin). A rotating adjustment screw is used to change the mercury level. The tip of an ivory pointer on the top of the mercury cistern indicates the zero of the scale. When the level of the mercury touches the tip, the atmospheric pressure is read at the top of the mercury column. For precis measurement, a Vernier scale is also attached to the main scale.

Atmospheric Pressure is estimated from the Main Scale Reading (MSR) and Vernier Scale Reading (VSR) as:

$$Atmosphere\ pressure = MSR + VSR \times Vernier\ constant$$

Correction of Barometer Readings

The reading of a mercury barometer reading should be corrected to the standard condition. Standard condition is defined as a temperature of 0 °C, where the density of mercury is 13.5951 g/cm^3 and a gravity acceleration of 980.665 cm/s^2.

The actual observation therefore needs to be corrected for the index error, temperature correction, and gravity acceleration as follows.

Corrections on Index Error

Index Error refers to the difference between the value indicated by an individual instrument and that of the standard Individual mercury barometers include index errors. The index error is found by comparison with the standard, and the value is stated on a "*comparison certificate*".

Corrections for Temperature

The temperature correction means to correct a barometric reading, obtained at a certain temperature, to a value when mercury and graduation temperatures are 0 °C. The temperature of the attached thermometer is used for this purpose.

The correction value for temperature C_t is expressed as follows:

$$C_t = -H \frac{(\mu - \lambda)t}{1 + \mu t}$$

Fig. 11.2 Fortin barometer

where:

H (hPa)   The barometric reading after the correction for index error.
t (°C)   The temperature indicated by the attached thermometer.
μ The volume expansion coefficient of mercury.
λ The linear expansion coefficient of the tube.

The values for correction at temperatures above 0 °C are negative and those below 0 °C are positive.

Corrections for Gravity

Gravity affects the height of the mercury column. After the corrections for index error and temperature, the reading under the local acceleration of gravity has to be reduced to the one under the standard gravity acceleration. This is called corrections for gravity. The gravity value for correction C_g is derived by:

$$C_g = H_0 - H = H\frac{g - g_0}{g_0}$$

where:

g_0 The standard gravity acceleration.
g The gravity acceleration at an observing point.
H The barometric reading after the index error and temperature corrections.
H_0 The value already corrected for gravitation.

The gravity acceleration used in corrections for gravity value is calculated to the fifth decimal place, in m/s². When the gravity acceleration at the observing point is larger than the standard gravity acceleration, the gravity value for correction is positive. Otherwise, the value for correction is negative.

Aneroid Barometer

An aneroid barometer does not contain any liquid. The aneroid barometer consists of barometer capsule having a spring to prevent it from being collapsed by the atmospheric pressure. The gears and levers are responsible for intensification and transmission of small amount of variations. The pressure reading is given by a pointer (Fig. 11.3).

Aneroid Barograph

The principle of the aneroid barograph is the same as that of the aneroid barometer, except that it uses a recording pen instead of the index needle. The change in atmospheric pressure causes displacement of the barometer capsule, which is transmitted to the recording pen through a reed and a lever (Fig. 11.4). The recording pen moves up and down on the side of the clock with recording drum to record the changes in atmospheric pressure.

Fig. 11.3 Aneroid barometer

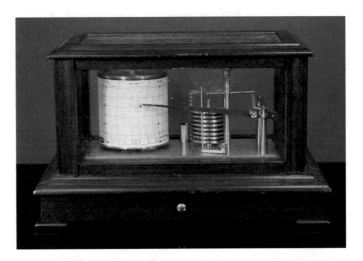

Fig. 11.4 Aneroid barograph

11.4 Atmospheric Pressure and Weather

The changes in weather are closely related to pressure variations as follows:

a. Falling barometer indicates rain or storm (bad weather).
b. Rising barometer indicates fair weather (clear and stable).
c. Steady barometer indicates steady or settled weather.
d. A continually rising pressure indicates occurrence of unsettled and cloudy weather.

Chapter 12
Automatic Weather Station

Abstract An Automatic Weather Station (AWS) is defined as a facility that automatically transmits or records observations obtained from measuring instruments. In an AWS, the measurements of meteorological elements are converted into electrical signals through sensors. The signals are then processed and transformed into meteorological data. The resulting information is finally transmitted the by wire or radio or automatically stored it on a recording medium. In this chapter, a brief introduction to AWS has been given. The chapter lists advantages of AWS over conventional methods. Various sensors used in Automatic Weather Stations are also described.

Keywords Automatic weather station · Site selection · Installation Sensors

12.1 Introduction

An Automatic Weather Station (AWS) is defined as a facility that automatically transmits or records observations obtained from measuring instruments. In an AWS, the measurements of meteorological elements are converted into electrical signals through sensors. The signals are then processed and transformed into meteorological data. The resulting information is finally transmitted the by wire or radio or automatically stored it on a recording medium. AWSs can be divided into real-time stations, which automatically transmit observed data at fixed times, and off-line stations, which record data on storage devices.

© Springer International Publishing AG 2017
L. Ahmad et al., *Experimental Agrometeorology: A Practical Manual*,
https://doi.org/10.1007/978-3-319-69185-5_12

12.2   Advantages of AWS

The advantages of using AWS can be summarized as:

- Continuous observation is possible.
- Observational data at manned stations can be obtained even when no staff are present.
- Fully automated systems can also be installed at inaccessible sites.
- Reduces observer numbers and operating costs.
- Since meteorological data are taken as electrical signals, observer errors in reading are eliminated.
- Standardized observation techniques enable the homogenization of observed data in regions where automatic weather observation is adopted.
- New observation elements can be added relatively easily by installing new instruments.
- Optimal measuring instruments with the appropriate level of measurement accuracy for the required observation can be chosen, and the need for observer training is eliminated.

12.3   Site Selection for AWS

The following points should be taken into account when selecting site for AWS:

- The site should meet the same meteorological requirements as those for conventional observations,
- Security measures against severe natural conditions and other interferences, including theft and vandalism, should be considered.

12.4   Installation and Construction of AWS

The AWS is generally installed on a plot measuring 10×12 m having adequate exposure for the sensors. The automatic weather station consists of the following main units (Fig. 12.1):

  i. Data logger field unit.
 ii. Data logger terminal.
iii. Data storage pack.
 iv. Solar panels.

Data Logger Field Unit

This unit collects and stores the weather data in the field. The data logger collects information from every sensor (temperature, RH, wind speed, wind direction, rainfall, solar radiation, etc.) and stores it either in its own memory or a memory card. It also processes most of the meteorological data i.e. provides average, minimum and maximum values of the recorded parameters.

Data Logger Terminal

The instructions to the field unit are given through the terminal, in other words, the terminal instructs the field unit about which sensor to use, what channel these sensors are wired to, when to store data, and how to label and organize data. It also helps in viewing the data on the display, check battery life and remaining memory on the data storage pack.

Data Storage Pack

The tabulated data is stored in the practical units of scientific measure using the data storage pack (DSP). For unloading the stored data, the data storage module can be connected to the data logger field unit. This unloaded data can then be transferred to a personal computer and stored on the computer hard disk for permanent record and further analysis.

Solar Panels

Solar panels provide the power to run the weather stations. The solar panels are equipped with rechargeable lead-acid batteries.

12.5 Sensors Used in Automatic Weather Station

Temperature and Relative Humidity Sensor

The model temperature and RH probe and is housed in plate gill radiation shield with a five feet lead length. This shield helps to eliminate radiation loading the sensor and also allows ventilation. RH and temperature probe sensors have $\pm 1\%$ and ± 0.3 °C accuracy at 20 °C. Measurement range for RH sensor is 0–100%.

Temperature Sensor

The temperature probe is used to measure air temperature. Temperature is sensed by thermistor which is extremely sensitive and exhibits a large resistance change with small change in temperature. Soil temperature is measured by another temperature probe which is electrically identical to air temperature probe, but is physically more rugged for burial applications.

Fig. 12.1 Automatic weather
station

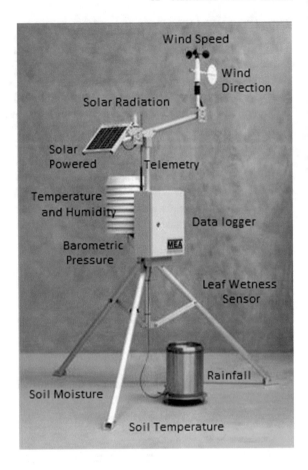

Wind Direction Sensor

The wind vane measures wind direction from 0° to 360° with a 5° accuracy. The sensor utilizes a potentiometer to vary the sensor resistance in relation to wind direction.

Wind Speed Sensor

The anemometer measures wind speed in the range of 0–45 m s^{-1} (0–160 km h^{-1}). This sensor is a three cup wheel assembly utilizing a magnet activated reed switch whose frequency is proportional to wind speed.

Radiation Sensor

This sensor is designed for field measurement of sun and sky radiation. The silicon pyranometer puts out a current which is dependent upon the solar radiation incident upon the sensor. The current is measured as the voltage drop across a fixed resistor.

Rainfall Sensor

This is a small adaptation of the standard Weather Bureau tipping bucket rain gauge. It measures rainfall at rates 50 mm per hour with an accuracy of ±1%. It is designed such that one alternate tip of the bucket occurs for each 0.25 mm of rainfall. Each tip actuates a magnetic switch. The rain gauge should be mounted on a level ground and at least 30 cm above the ground surface.

Besides these major field components, the whole system also consists of the following major sub-systems.

a. Data Relay transponder on board the satellite (KALPANA-1, INSAT 3A).
b. Data receiving and dissemination system.

The dependability on AWS has increased considerably and the capabilities of AWS provide an additional and reliable tool for weather forecasting and monitoring.

Chapter 13
Estimation of Climate Change Through Trend Analysis

Abstract Climate change effects on the environment and humankind. The chapter describes weather climate and the difference between them. Some emphasis has been laid on describing climate change and climate variability. The chapter also includes information for studying Temperature and rainfall Variation and Trends.

Keywords Climate change · Trend analysis · Mann-Kendall test Examples

13.1 Weather and Climate

Weather refers to atmospheric conditions that occur locally over short periods of time—from minutes to hours or days. Familiar examples include rain, snow, clouds, winds, floods or thunderstorms. Climate, on the other hand, refers to the long-term regional or even global average of temperature, humidity and rainfall patterns over seasons, years or decades.

13.2 Global Warming

Global warming is the term used to describe the upward temperature trend of the Earth. The average surface temperature has gone up by about 0.8 °C since 1880 and the phenomenon has been most noticeable since late 1970s.

13.3 Global Warming and Climate Change

Global warming causes climate change, so the two terms are very much related. Global warming is the term used to describe the current increase in the Earth's average temperature. Climate change refers not only to global changes in temperature

© Springer International Publishing AG 2017
L. Ahmad et al., *Experimental Agrometeorology: A Practical Manual*,
https://doi.org/10.1007/978-3-319-69185-5_13

but also to changes in wind, precipitation, the length of seasons as well as the strength and frequency of extreme weather events like droughts and floods.

Another difference between the two terms is that global warming is a worldwide phenomenon while climate change can be seen at global, regional or even more local scales.

13.4 Climate Change

Climate change is a long-term continuous change (increase or decrease) to average weather conditions (e.g. average temperature) or the range of weather (e.g. more frequent and severe extreme storms). Long-term refers to decades or longer. Climate change is defined by United Nations Framework Convention on Climate Change (UNFCCC) as:

> A change of climate which is attributed directly or indirectly to human activity that alters the composition of the global atmosphere and which is in addition to natural climate variability observed over comparable time periods.

Climate change is a slow and gradual process and is very difficult to perceive without scientific records.

13.5 Climate Variability

The fluctuation of climate above or below a long-time average value is climate variability. The term "Climate Variability" is often used to denote deviations of climatic statistics over a given period of time (e.g. a month, season or year) when compared to long-term statistics for the same calendar period. Climate variability is measured by these deviations, which are usually termed anomalies.

13.6 Climate Variability and Climate Change

Climate variability is concerned with the changes that occur within smaller time-frames, such as a month, a season or a year. On the other hand, climate change considers changes that occur over a longer period of time, typically over decades or longer. In climate variability there is persistence of "anomalous" conditions—when events that used to be rare occur more frequently, or vice versa.

13.7 Signal and Noise

The year-to-year variations are often referred to as noise, whereas the trend is the signal.

13.8 Variation and Trends in Temperature and Rainfall

In order to detect trends in temperature values, several statistical methods have been put to practice. Two of the most common methods of detecting trends in temperature data are linear regression test and while Mann-Kendall test.

Regression Test for Linear Trend

A straight line is fitted to the data and statistically tested if the slope is different from zero or not. A straight line of the form $y = a + bx$ is fitted to the data. The following statistics are then computed:

$$a = \bar{Y} - b\bar{X}$$

$$b = \frac{\sum (X_i - \bar{X})(Y_i - \bar{Y})}{\sum (X_i - \bar{X})^2}$$

The test statistic, t is given as:

$$t = \frac{b}{S_b}$$

For which,

$$S_b^2 = \frac{S^2}{\sum (X_i - \bar{X})^2}$$

$$S = \left[\frac{\sum \varepsilon_i^2}{(n - 2)} \right]^{1/2}$$

$$\sum \varepsilon_i^2 = \sum (Y_i - \bar{Y})^2 - b^2 \sum (X_i - \bar{X})^2$$

where,

$\bar{X}$ Mean of values of X,
$\bar{Y}$ Mean of values of Y,
S_b Standard error of b,
ε_i^2 Sum of squares of residuals.

The test statistic 't' is tested using student's t-test.

The slope is significant if $|t| > t_{standard}$.

Where $t_{standard} = t_{1-\alpha/2,\,(n-2)}$ and α is the level of significance level. The values of $t_{standard}$ at different levels of significance are:

at 10% significance level = 1.73
at 5% significance level = 2.10
at 1% significance level = 2.88.

Mann-Kendall Test

The Mann-Kendall statistic S of a series x is calculated as (Mann 1945; Kendall 1975):

$$S = \sum_{i=1}^{n-1} \sum_{j=i+1}^{n} sgn(x_j - x_i)$$

where,

$$sgn(x_j - x_i) = f(x) = \begin{cases} 1 & if & (x_j > x_i) \\ 0 & if & (x_j = x_i) \\ -1 & if & (x_j < x_i) \end{cases}$$

The variance associated with the test static S is calculated as:

$$Var(S) = \frac{n(n-1)(2n+5) - \sum_{k=1}^{m} t_k(t_k - 1)(2t_k + 5)}{18}$$

where,

m No. of tied groups.
t_k No. of data points in group k.

If the sample size is more than 10 (n > 10), the test statistic is calculated as:

$$Z_{MK} = \begin{cases} \frac{S-1}{\sqrt{Var(S)}} & if & S > 0 \\ 0 & if & S = 0 \\ \frac{S+1}{\sqrt{Var(S)}} & if & S < 0 \end{cases}$$

A trends is considered significant if $|Z(S)|$ is more than the standard normal deviate $(Z_{(1-\alpha/2)})$ for the chosen value of α. α is the significance level of the test.

For trends to be significant at:

1% $|Z_{MK}| > 1.645$
5% $|Z_{MK}| > 1.96$
10% $|Z_{MK}| > 2.576$

Example 13.1 The annual rainfall at a station is given. Find the presence of trend by linear regression test at 1% significance level.

Year	1	2	3	4	5	6	7	8	9	10
Rainfall (cm)	90	100	80	105	85	84	110	82	113	86

Solution:

For this case:

$\bar{X} = 5.5$
$\bar{Y} = 93.5$
$\sum(X_i - \bar{X})^2 = 82.5$
$\sum(Y_i - \bar{Y})^2 = 1372.5$
$\sum(Y_i - \bar{Y})(X_i - \bar{X}) = 39.5$
$b = 0.478$
$a = 90.871$
$\sum e_i^2 = 1353.65$
$S = 13$
$S_b = 2.048$
$t = 0.233$
$t_{0.995,8} = 3.36$
$|t| < 3.36$, there is no trend at 1% significance level

Example 13.2 Find the presence of trend for the data given in example 13.1 by Mann-Kendall test at 10% significance level.

Solution:

The calculations for Mann-Kendall test are given:
The value of S is given as:

$$S = \sum_{i=1}^{n-1} \sum_{j=i+1}^{n} sgn(x_j - x_i)$$

The calculation of $sgn(x_j - x_i)$ is given in Table.

Year	Rainfall (cm)										
		90	100	80	105	85	84	110	82	113	86
1	90										
2	100	1									

(continued)

(continued)

Year	Rainfall (cm)											
3	80	−1	−1									
4	105	1	1	1								
5	85	−1	−1	1	−1							
6	84	−1	−1	1	−1	−1						
7	110	1	1	1	1	1	1					
8	82	−1	−1	1	−1	−1	−1	−1				
9	113	1	1	1	1	1	1	1	1			Sum
10	86	−1	−1	1	−1	1	1	−1	1	−1		5

$$S = 5$$

Now,

$$Var(S) = \frac{n(n-1)(2n+5) - \sum_{k=1}^{m} t_k(t_k-1)(2t_k+5)}{18}$$

No. of ties = 0

$$Var(S) = \frac{n(n-1)(2n+5)}{18}$$

$$Var(S) = \frac{10(10-1)(2 \times 10 + 5)}{18}$$

$$Var(S) = 125$$

$$Z_{MK} = \frac{S-1}{\sqrt{Var(S)}}$$

$$Z_{MK} = \frac{5-1}{\sqrt{125}}$$

$$Z_{MK} = 0.36$$

Since, $|Z_{MK}| < 2.576$

Therefore the data is trend free at 10% significance level.

Chapter 14
Growing Degree Days to Forecast Crop Stages

Abstract Plant development depends on temperature. Plants require a specific amount of heat to develop from one point in their life-cycle to another. The ability to predict a specific crop stage, relative to insect and weed cycles, permits better management. This is especially important when three or more crops are being grown on the same farm, each with a different management schedule for pesticide application, fertility management and harvest. The chapter introduces the concept of growing degree days, and other thermal indices used to study and monitor plant growth.

Keywords GDD · Crop forecast · Units · Limitation

14.1 Introduction

Plant development depends on temperature. Plants require a specific amount of heat to develop from one point in their life-cycle to another. The ability to predict a specific crop stage, relative to insect and weed cycles, permits better management. This is especially important when three or more crops are being grown on the same farm, each with a different management schedule for pesticide application, fertility management and harvest.

14.2 Heat Unit or Growing Degree Days

The heat unit or GDD concept was proposed to explain the relationship between growth duration and temperature. The concept assumes a direct and linear relationship between growth and temperature. GDD is the departure from the mean daily temperature above the minimum threshold or base temperature.

GDD are calculated by determining the mean daily temperature and subtracting it from the base temperature needed for growth of the organism. The GDD value for one day is represented by the following equation:

$$GDD = \frac{(T_{max} + T_{min})}{2} - T_b$$

where,

T_{max} Daily Maximum Temperature
T_{min} Daily Minimum Temperature
T_b Base Temperature.

The temperature below which the plant (organism) does not grow or grows very slowly. At temperatures above the minimum, plant development growth rate increases as temperature increases up to optimum. While there is only limited plant growth at temperatures slightly above freezing, germination and seedling growth do not occur at temperatures between 0 and 5 °C. The base temperature varies from crop to crop, stage and seasonal conditions of the crop. The base temperatures of different crops are:

Rice = 10	Wheat = 4.5	Maize = 10	Pearl Millet = 10	Oats = 4.5

14.3 Limitations of Growing Degree Day Concept

1. It assumes linear relationship between temperature and development while the actual relationship is curvilinear.
2. It gives more importance to higher temperatures although it is detrimental to growth.
3. Diurnal variation is not taken into consideration though it has considerable influence on growth.
4. No allowance is made for base temperature changes with advancing stage of crop development.

14.4 Modification of GDD Expression

Some modifications have been suggested by different scientists to overcome the limitations of GDD concept.

Method I

When $T_{av} < T_{base}$, then $T_{av} = T_{base}$
When $T_{av} > T_{UT}$, then $T_{av} = T_{UT}$

Method II

When T_{max} or $T_{min} < T_{base}$, then T_{max} or $T_{min} = T_{base}$
When T_{max} or $T_{min} > T_{UT}$, then T_{max} or $T_{min} = T_{UT}$

where,

T_{av} Average Temperature
T_{base} Base Temperature
T_{max} Daily Maximum Temperature
T_{min} Daily Minimum Temperature
T_{UT} Upper Threshold Temperature.

14.5   Photo-Thermal Unit (PTU) Helio-Thermal Unit (HTU) and Hydro-Thermal Unit (HYTU)

PTU is the product of GDD and day length on any day (Nuttonson 1948). It is calculated as:

$$PTU = GDD \times Day\ Length$$

HTU is the product of GDD and actual Bright Sunshine Hours in any day.

$$HTU = GDD \times No.\ of\ Actual\ Sunshine\ Hours$$

HYTU is the product of GDD and average relative humidity for that day.

$$HYTU = GDD \times Av.\ RH\ (\%)$$

14.6   Heat Use Efficiency (HUE), Photo-Thermal Use Efficiency (PTUE) and Helio-Thermal Use Efficiency (HTUE)

Heat Use Efficiency (HUE) is defined as the biomass accumulation during a given period per day. It is computed in kg ha^{-1} C^{-1}. It is calculated as:

$$HUE \, (\text{kg ha}^{-1} \, {}^{\circ}\text{C}^{-1} \, \text{day}) = \frac{Seed \, yield(\text{kg ha}^{-1})}{Accumulated \, Heat \, Units({}^{\circ}C)}$$

PTUE and HTUE are used to compare the relative performance of the crops under different treatments. These are calculated as:

$$PTUE(\text{kg ha}^{-1} \, {}^{\circ}\text{C}^{-1} \, \text{day}) = \frac{Seed \, yield \, (\text{kg ha}^{-1})}{PTU({}^{\circ}C)}$$

$$HTUE \, (\text{kg ha}^{-1} \, {}^{\circ}\text{C}^{-1} \, \text{day}) = \frac{Seed \, yield \, (\text{kg ha}^{-1})}{HTU({}^{\circ}C)}$$

14.7   Photo-Thermal Index (PTI)

Photo-thermal Index (PTI) expressed as degree days per growth day for different growth stage and whole duration of the crop will be calculated using the formula (Sastry and Chakravarty 1982):

$$PTI = \frac{GDD}{No. \, of \, days \, taken \, between \, two \, phenophases}$$

Chapter 15
Agro-climatic and Agro-ecological Zones of India

Abstract India exhibits a variety of land scopes and climatic conditions those are reflected in the evolution of different soils and vegetation. These also exists a significant relationship among the soils, land form climate and vegetation. Regions are delineated such that each one is as uniform as possible with respect to physiographic, climate, length of growing period (LGP) and soils for macro level and land use planning and effective transfer of agro-technology. Various Agro Climatic Zones of India according to the planning commission of India have been explained along with the classification by Indian Council of Agricultural Research. A section on Agro-ecological Zones of India has been included. The chapter also introduces the Agro-Climatic zones of Jammu and Kashmir and the Micro Agro-Climatic zones of the state. The State of Jammu and Kashmir is located almost in the middle of three climatic regimes of Asia. The chapter introduces the Agro-Climatic zones of Jammu and Kashmir based on physiography. Each province of the state of Jammu and Kashmir has been further delineated into micro agro-climatic zones which have been discussed.

Keywords Agroclimatic zones · India · Jammu and Kashmir · Case study

15.1 Introduction

The important rational planning for effective land use to promote efficient is well recognized. The ever increasing need for food to support growing population @ 2.1% (1860 millions) in the country demand a systematic appraisal of our soil and climatic resources to recast effective land use plan. Since the soils and climatic conditions of a region largely determine the cropping pattern and crop yields. Reliable information on agro ecological regions homogeneity in soil site conditions is the basic to maximize agricultural production on sustainable basis. This kind of systematic approach may help the country in planning and optimizing land use and preserving soils, environment. India exhibits a variety of land scopes and climatic conditions those are reflected in the evolution of different soils and vegetation.

© Springer International Publishing AG 2017
L. Ahmad et al., *Experimental Agrometeorology: A Practical Manual*,
https://doi.org/10.1007/978-3-319-69185-5_15

These also exists a significant relationship among the soils, land form climate and vegetation. Regions are delineated such that each one is as uniform as possible with respect to physiographic, climate, length of growing period (LGP) and soils for macro level and land use planning and effective transfer of agro- technology.

15.2   Agro Climatic Zones

An "Agro-climatic zone" is a land unit in terms of major climates, suitable for a certain range of crops and cultivars. The planning aims at scientific management of regional resources to meet the food, fiber, fodder and fuel wood without adversely affecting the status of natural resources and environment. Agro-climatic conditions mainly refer to soil types, rainfall, temperature and water availability which influences the type of vegetation. An agro-ecological zone is the land unit carved out of agro-climatic zone superimposed on landform which acts as modifier to climate and length of growing period.

India has been divided into 24 agro-climatic zones by Krishnan and Mukhtar Sing, in 1972 by using "Thornthwait indices" The planning commission. as a result of mid-term appraisal of planning targets of VII plan (1985–90) divided the country into 15 broad agro-climatic zones based on physiographic and climate (Fig. 15.1). These are further divided into 72 sub-zones taking into account the physical attributes and socio-economic conditions of the region. The emphasis was given on the development of resources and their optimum utilization in a suitable manner with in the frame work of resource constraint and potential of each region (Khanna 1989).

Western Himalayan Region

The Western Himalayan Region covers Jammu and Kashmir, Himachal Pradesh and the hill region of Uttarakhand. Topography and temperatures show great variation. Average temperature in July ranges between 5 and 30 °C, while in January it ranges between 5 and −5 °C. Mean annual rainfall varies between 75 cm to 150 cm; in Ladakh, however, it is less than 30 cm. There is alluvial soil in the valleys of Kashmir, Kullu and Dun, and brown soil in the hills. The valley floors grow rice, while the hilly tracts grow maize in the kharif season. Winter crops are barley, oats, and wheat. The region supports horticulture, especially apple orchards and other temperate fruits such as peaches, apricot, pears, cherry, almond, litchis, walnut, etc. Saffron is grown in this region. The high altitude alpine pastures, locally known as 'dhoks' or 'margs', are used by the Gujjars, Bakarwals and Gaddis to rear their sheep, goats, cattle and horses. The economy is largely agrarian. The main problems of this region are poor accessibility, soil erosion, landslides, inclement weather, inadequacy of marketing and storage facilities. The population is generally rural-based and poor. Research in better seeds and extension service for agricultural development are required.

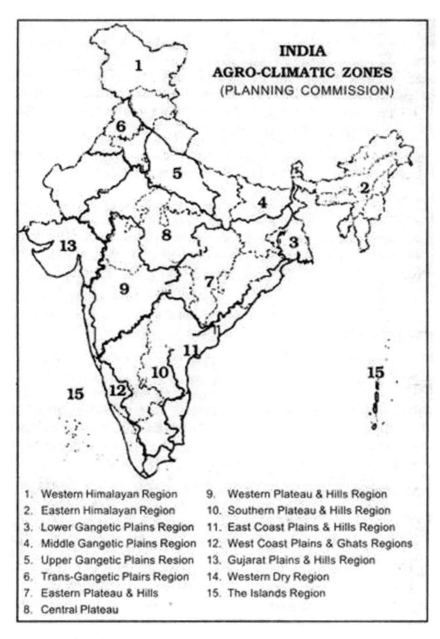

Fig. 15.1 Agro-climatic zones of India (planning commission)

Eastern Himalayan Region

The Eastern Himalayan Region includes Arunachal Pradesh, the hills of Assam, Sikkim, Meghalaya, Nagaland, Manipur, Mizoram, Tripura, and the Darjeeling district of West Bengal. The topography is rugged. Temperature variation is between 25 and 30 °C in July and between 10 and 20 °C in January. Average rainfall is between 200 and 400 cm. The red-brown soil is not highly productive Jhuming (shifting cultivation) prevails in the hilly areas.

The main crops are rice, maize, potato, tea. There are orchards of pineapple, litchi, oranges and lime. Infrastructural facilities in the region need to be improved and shifting cultivation controlled by developing terrace farming.

Lower Gangetic Plain Region

West Bengal (except the hilly areas), eastern Bihar and the Brahmaputra valley lie in this region. Average annual rainfall lies between 100 and 200 cm. Temperature in July varies from 26 to 41 °C and for January from 9 to 24 °C. The region has adequate storage of ground water with high water table. Rice is the main crop which at times yields three successive crops (Aman, Aus and Boro) in a year. Jute, maize, potato, and pulses are other important crops. Planning strategies include improvement in rice farming, horticulture (banana, mango and citrus fruits), pisciculture, poultry, livestock, forage production and seed supply.

Middle-Gangetic Plain Region

The Middle-Gangetic Plain region includes large parts of Uttar Pradesh and Bihar. The average temperature in July varies from 26 to 41 °C and that of January 9 to 24 °C average annual rainfall is between 100 and 200 cm. It is a fertile alluvial plain drained by the Ganga and its tributaries. Rice, maize, millets in kharif, wheat, gram, barley, peas, mustard and potato in rabi are important crops. Alternative farming systems and utilizing chaur lands for pisciculture are some measures to boost agricultural production. Reclamation of user lands, wastelands, and fallow lands for agriculture and allied activities (agro-forestry, silviculture, floriculture etc.) should be done.

Upper Gangetic Plains Region

In the Upper Gangetic Plains region come the central and western parts of Uttar Pradesh and the Hardwar and Udham Nagar districts of Uttarakhand. The climate is sub-humid continental with temperature in July between 26 and 41 °C and temperature in January between 7 and 23 °C. Average annual rainfall is between 75 and 150 cm. The soil is sandy loam. Canal, tube-well and wells are the main source of irrigation. This is an intensive agricultural region wherein wheat, rice, sugarcane, millets, maize, gram, barley, oilseeds, pulses and cotton are the main crops. Besides modernizing traditional agriculture, the region needs special focus on dairy development and horticulture. Strategies should include developing multiple mixed cropping patterns.

Trans-Ganga Plains Region

This region (also called the Satluj-Yamuna Plains) extends over Punjab, Haryana, Chandigarh, Delhi and the Ganganagar district of Rajasthan. Semi-arid characteristics prevail over the region, with July's mean monthly temperature between 25 and 40 °C and that of January between 10 and 20 °C. The average annual rainfall varies between 65 and 125 cm. The soil is alluvial which is highly productive. Canals and tube-wells and pumping sets have been installed by the cultivators and the governments. The intensity of agriculture is the highest in the country. Important crops include wheat, sugarcane, cotton, rice, gram, maize, millets, pulses and oilseeds etc. The region has the credit of introducing Green Revolution in the country and has adopted modern methods of farming with greater degree of mechanization. The region is also facing the menace of water logging, salinity, alkalinity, soil erosion and falling water table.

Some steps that may be required to make agriculture in the region more sustainable and productive are:

i. diversion of some rice-wheat area to other crops like maize, pulses, oilseeds and fodder;
ii. development of genotypes of rice, maize and wheat with inbuilt resistance to pests and diseases;
iii. promotion of horticulture besides pulses like tur and peas in upland conditions;
iv. cultivation of vegetables in the vicinity of industrial clusters;
v. supply of quality seeds of vegetables and planting material for horticulture crops;
vi. development of infra-structure of transit godowns and processing to handle additional fruit and vegetable production;
vii. implementation of policy and programs to increase productivity of milk and wool; and
viii. Development of high quality fodder crops and animal feed by stepping up area under fodder production.

Eastern Plateau and Hills

This region includes the Chotanagpur Plateau, extending over Jharkhand, Orissa, Chhattisgarh and Dandakaranya. The region enjoys 26–34 °C of temperature in July, 10–27 °C in January and 80–150 cm of annual rainfall. Soils are red and yellow with occasional patches of laterites and alluviums. The region is deficient in water resources due to plateau structure and non-perennial streams. Rain-fed agriculture is practiced growing crops like rice, millets, maize, oilseeds, ragi, gram and potato. Steps to improve agricultural productivity and income include cultivation of high value crops of pulses like tur, groundnut and soyabean etc. on upland rain-fed areas, growing crops like urad, castor, and groundnut in kharif and mustard and vegetables in irrigated areas, improvement of indigenous breeds of cattle and buffaloes, extension of fruit plantations, renovation including desilting of existing tanks and excavation of new tanks, 95.32 lakh ha of acidic lands through lime treatment, development of inland fisheries in permanent water bodies, and adopting integrated watershed development approach to conserve soil and rain water.

Central Plateau and Hills

The region is spread over Bundelkhand, Baghelkhand, Bhander Plateau, Malwa Plateau, and Vindhyachal Hills. Semi-arid climatic conditions prevail over the region with temperature in July 26–40 °C, in January 7–24 °C and average annual rainfall from 50 to 100 cm. Soils are mixed red, yellow and black. There is scarcity of water. Crops grown are millets, wheat, gram, oilseeds, cotton and sunflower. In order to improve agricultural returns, measures to be adopted are water conservation through water saving devices like sprinklers and drip system; dairy development, crop diversification, ground water development, reclamation of ravine lands.

Western Plateau and Hills

Comprising southern part of Malwa plateau and Deccan plateau (Maharashtra), this is a region of the regur (black) soil with July temperature between 24 and 41 °C, January temperature between 6 and 23 °C and average annual rainfall of 25–75 cm. Wheat, gram, millets, cotton, pulses, groundnut, and oilseeds are the main crops in the rain-fed areas, while in the irrigated areas, sugarcane, rice, and wheat, are cultivated. Also grown are oranges, grapes and bananas. Attention should be paid to increasing water efficiency by popularizing water saving devices like sprinklers and drip system. The lower value crops of jowar, bajra and rainfed wheat should give way to high value oilseeds. Five per cent area under rain-fed cotton and jowar could be substituted with fruits like ber, pomegranate, mango and guava. Improvement of milk production of cattle and buffalo through cross-breeding along with poultry development should be encouraged.

Southern Plateau and Hills

This region falls in interior Deccan and includes parts of southern Maharashtra, the greater parts of Karnataka, Andhra Pradesh, and Tamil Nadu uplands from Adilabad District in the north to Madurai District in the south. The mean monthly temperature of July varies between 25 and 40 °C, and the mean January temperature is between 10 and 20 °C. Annual rainfall is between 50 and 100 cm. It is an area of dry-zone agriculture where millets, oilseeds, and pulses are grown. Coffee, tea, cardamom and spices are grown along the hilly slopes of Karnataka plateau. Some of the area under coarse cereals may be diverted to pulses and oilseeds. Horticulture, dairy development and poultry farming should be encouraged.

Eastern Coastal Plains and Hills

In this region are the Coromandal and northern Circar coasts of Andhra Pradesh and Orissa. The mean July temperature ranges between 25 and 35 °C and the mean

January temperature varies between 20 and 30 °C. The mean annual rainfall varies between 75 and 150 cm. The soils are alluvial, loam and clay and are troubled by the problem of alkalinity. Main crops include rice, jute, tobacco, sugarcane, maize, millets, groundnut and oilseeds. Main agricultural strategies include improvement in the cultivation of spices (pepper and cardamom) and development of fisheries. These involve increasing cropping intensity using water-efficient crops on residual moisture, discouraging growing of rice on marginal lands and bringing such lands under alternate crops like oilseeds and pulses; diversifying cropping and avoiding mono-cropping; developing horticulture in upland areas, social forestry and dairy-farming.

Western Coastal Plains and Ghats

Extending over the Malabar and Konkan coastal plains and the Sahyadris, the region is humid with the mean July temperature varying between 25 and 30 °C and mean January temperatures between 18 and 30 °C. The mean annual rainfall is more than 200 cm. The soils are laterite and coastal alluvial. Rice, coconut, oilseeds, sugarcane, millets, pulses and cotton are the main crops. The region is also famous for plantation crops and spices which are raised along the hill slopes of the Western Ghats. The agricultural development must focus attention on raising of high value crops (pulses, spices, and coconut). Development of infra- structural facilities and promotion to prawn culture in brackish water should be encouraged.

Gujarat Plains and Hills

This region includes the hills and plains of Kathiawar, and the fertile valleys of Mahi and Sabarmati rivers. It is an arid and semi-arid region with the mean July temperature reading 30 °C and that of January about 25 °C. The mean annual rainfall varies between 50 and 100 cm. Soils are regur in the plateau region, alluvium in the coastal plains, and red and yellow soils in Jamnagar area. Groundnut, cotton, rice, millets, oilseeds, wheat and tobacco are the main crops. It is an important oilseed producing region.

The main strategy of development in this region should be canal and groundwater management, rain water harvesting and management, dry land farming, agro-forestry development, wasteland development and developing marine fishing and brackish/back-water aquaculture development in coastal zones and river deltas.

Western Dry Region

Extending over Rajasthan, West of the Aravallis, this region has an erratic rainfall of an annual average of less than 25 cm. The desert climate further causes high evaporation and contrasting temperatures—28 to 45 °C in June and 5 to 22 °C in January. Bajra, jowar, and moth are main crops of kharif and wheat and gram in rabi. Livestock contributes greatly in desert ecology. The main areas needing a thrust for development are rainwater harvesting, increasing yield level of horticultural crops like water melon, guava and date palm, adopting high quality germ- plasm in cattle to improve their breed; and adopting silvi-pastoral system over wastelands.

Island Region

The island region includes Andaman-Nicobar and Lakshadweep which have typically equatorial climate (annual rainfall less than 300 cm; the mean July and January temperature of Port Blair being 30 and 25 °C respectively). The soils vary from sandy along the coast to clayey loam in valleys and lower slopes. The main crops are rice, maize, millets, pulses, arecanut, turmeric and cassava. Nearly half of the cropped area is under coconut. The area is covered with thick forests and agriculture is in backward stage. The main thrust in development should be on crop improvement, water management and fisheries. Improved variety of rice seeds should be popularized so as to enable farmers to take two crops of rice in place of one. For fisheries development multi-purpose fishing vessels for deep sea fishing should be introduced, suitable infrastructure for storage and processing of fish should be built up, and brackish water prawn culture should be promoted in the coastal areas.

15.3   Classification by ICAR

The State Agricultural Universities have divided each state into sub-zones, under the National Agricultural Research Project (NARP) of ICAR. Based on the rainfall pattern, cropping pattern and administrative units, 127 agro-climatic zones are indicated. The number of zones in each state are given in Table 15.1. The details of these zones are given in Appendix C (Fig. 15.2).

Table 15.1 No. of Agro-climatic zones in different states of India (NARP, ICAR)

State	No. of zones	State	No. of zones
Andhra Pradesh	7	Madhya Pradesh	12
Assam	6	Maharashtra	9
Bihar	6	N E Hill region	6
Gujarat	8	Orissa	10
Haryana	2	Punjab	5
Himachal Pradesh	4	Rajasthan	9
Jammu and Kashmir	5	Tamil Nadu	7
Karnataka	10	Uttar Pradesh	10
Kerala	5	West Bengal	6
Total 127			

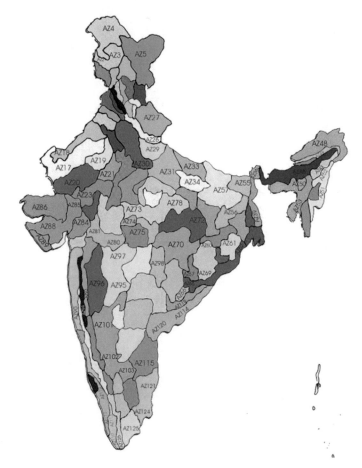

Fig. 15.2 Agro-climatic zones of India (NARP, ICAR)

15.4   Agro-ecological Zones of India

Based on physiographic features, soil characteristics, bio-climatic types (rainfall, potential evapotranspiration, soil storage) and length of the growing period, India is divided into 20 agro-ecological regions. The details are given in Table 15.2.

Table 15.2 Agro-ecological zones of India

AEZ no.	Agro-ecological region	Geographical area (million ha)	Gross cropped area (million ha)	Physiography	Precipitation (mm)	PET (mm)	Length of growing period (days)	Major crops
1	Cold arid eco region with shallow skeletal soils	15.2 (4.7%)	0.07	Western Himalayas	<150	<800	<90	Vegetables, millets, wheat, fodder, barley, pulses
2	Hot arid eco region with desert and saline soils	31.9	20.85	Western Plain and Kachchh Peninsula	<300	1500–2000	<90	Millets, fodder, pulses
3	Hot arid eco region with red and black soils	4.9	4.18	Deccan Plateau	400–500	1800–1900	<90	Sorghum, safflower, cotton, groundnut, sunflower, sugar cane
4	Hot semi-arid eco region with alluvium-derived soils	32.2	30.05	Northern Plain and Central Highlands including parts of Gujarat Plains	500–800	1400–1900	90–150	Millets, wheat, pulses, maize; irrigated cotton and sugar cane
5	Hot semi-arid eco region with medium and deep black soils	17.6	11.04	Central (Malwa) Highlands, Gujarat Plains and Kathiawar Peninsula	500–1000	1600–2000	90–150	Millets, wheat, pulses

(continued)

Table 15.2 (continued)

AEZ no.	Agro-ecological region	Geographical area (million ha)	Gross cropped area (million ha)	Physiography	Precipitation (mm)	PET (mm)	Length of growing period (days)	Major crops
6	Hot semi-arid eco region with shallow and medium (dominant) black soils	31	25.02	Deccan Plateau	600–1000	1600–1800	90–150	Millets, cotton, pulses, sugar cane under irrigation
7	Hot semi-arid eco region with red and black soils	16.5	6.19	Deccan (Telangana) Plateau and Eastern Ghats	600–1000	1600–1700	90–150	Millets, oilseeds, rice, cotton and sugar cane under irrigation
8	Hot semi-arid eco region with red loamy soils	19.1	6.96	Eastern Ghats (Tamil Nadu uplands) and Deccan Plateau (Karnataka)	600–1000	1300–1600	90–150	Millets, pulses, oilseeds (groundnut), sugar cane and rice under irrigation
9	Hot subhumid (dry) eco region with alluvium-derived soils	12.1	11.62	Northern Plain	1000–1200	1400–1800	150–180	Rice, wheat, pigeon pea, sugar cane, mustard, maize
10	Hot subhumid eco region with red and black soils	22.3	14.55	Central Highlands (Malwa and Bundelkhand)	1000–1500	1300–1500	150–180	Rice, wheat, sorghum, soybean, gram, pigeon pea
11	Hot subhumid eco region with red and yellow soils	11.1	6.47	Eastern Plateau (Chhattisgarh Region)	1200–1600	1400–1500	150–180	Rice, millets, wheat, pigeon pea, green gram, black gram

(continued)

Table 15.2 (continued)

AEZ no.	Agro-ecological region	Geographical area (million ha)	Gross cropped area (million ha)	Physiography	Precipitation (mm)	PET (mm)	Length of growing period (days)	Major crops
12	Hot subhumid eco region with red and lateritic soils	26.8	12.09	Eastern (Chhota Nagpur) Plateau and Eastern Ghats	1000–1600	1400–1700	150–180	Rice, pulses, millets
13	Hot subhumid (moist) eco region with alluvium-derived soils	11.1	10.95	Eastern Plains	1400–1600	1300–1500	180–210	Rice, wheat, sugar cane
14	Warm subhumid to humid with inclusion of perhumid eco region with brown forest and podzolic soils	18.2	3.2	Western Himalayas	1600–2000	800–1300	180–210	Wheat, millets, maize, rice
15	Hot subhumid (moist) to humid (inclusion of perhumid) eco regions with alluvial-derived soils	12.1	8.99	Bengal Basin and Assam Plain	1400–2000	1000–1400	>210	Rice, jute, plantation crops
16	Warm perhumid eco region with brown and red hill soils	9.6	1.37	Eastern Himalayas	2000–4000	<1000	>210	Rice, millets, potato, maize, sesame, Jhum* cultivation is common

(continued)

Table 15.2 (continued)

AEZ no.	Agro-ecological region	Geographical area (million ha)	Gross cropped area (million ha)	Physiography	Precipitation (mm)	PET (mm)	Length of growing period (days)	Major crops
17	Warm perhumid eco region with red and lateritic soils	10.6	1.56	North-Eastern Hills	1600–2600	1000–1100	> 210	Rice, millets, potato, plantation crops, Jhum* cultivation is common
18	Hot subhumid to semi-arid eco region with coastal alluvium-derived soils	8.5	6.12	Eastern Coastal Plains	900–1600	1200–1900	90 > 210	Rice, coconut, black gram, lentil, sunflower, groundnut
19	Hot humid perhumid eco region with red, lateritic and alluvium-derived soils	11.1	5.7	Western Ghats and Coastal Plains	2000–3200	1400–1600	>210	Rice, tapioca, coconut, spices
20	Hot humid/perhumid island eco region with red loamy and sandy soils	0.8	0.05	Islands of Andaman & Nicobar and Lakshadweep	1600–3000	1400–1600	>210	Rice, coconut, areca nut, oil palm

15.5 Agro-climatic Zones of Jammu and Kashmir—Case Study

The State of Jammu and Kashmir extends between 32° 17' and of 37° 5' North and 73° 26' and 80° 30' East. The State is located almost in the middle of three climatic regimes of Asia. In its south border lies the weak monsoon zone of Punjab. On the north-east the State is bordered by the vast arid plateau of Tibet while the North-west border areas face the eastern limits of Mediterranean climatic region. This geographical position, coupled with the varied physiography, provides the State a wide climatic variation.

The State has been divided into four broad macro-climatic zones (i) sub-tropical (ii) valley temperate (iii) intermediate (iv) cold-arid. The State has mostly a mountainous area and occupies a central position in the continent of Asia. Out of 3.5 million ha of mountainous area of India, nearly two third i.e. 2.3 million ha are found exclusively in Jammu and Kashmir State. The State is bounded on the north by Chinese and Russian territories, on the east by Tibet, on the south by Punjab (India) and on the West by Pakistan (Fig. 17.1). It has high mountainous terrains with many snow-covered peaks ranging in altitude from 554 to 7077 m amsl in North and North-west, which are succeeded towards the South by lower range of hills (Fig. 15.3).

Total geographical area of the State is 222,236 km^2 including 78,114 km^2 (35.15%) under Pakistan, and 42,735 km^2 (19.23%) under China. Ladakh is the largest hilly arid zone which occupies 58,321 km^2 (42.00%).

Agro-climatic Zones of Jammu and Kashmir

On the basis of physiography, the state may be divided into three main regions

  i. outer Himalayas which comprise of Jammu province,
 ii. lesser Himalayas which comprises of Kashmir Valley, and
iii. inner Himalayas which comprises of Ladakh province.

Corresponding to these regions, there are four major agro-climatic zones detailed in the state. These are described below.

Jammu Region

Jammu region comprises of two major agro-climatic zones viz. low altitude sub-tropical zone and mid to high intermediate zone.

Low Altitude Subtropical Zone (JK-1)

The zone is characterized by monsoon, concentration of precipitation, hot spell of summer, relatively dry but pronounced winter and prevalence of alluvial soils. It comprises of whole Jammu district and lower parts of Kathua, Udhampur, Poonch and Rajouri districts. Maximum rainfall is received during July–September. The mean height above sea level ranges from less than 300 m to nearly 1350 m. May,

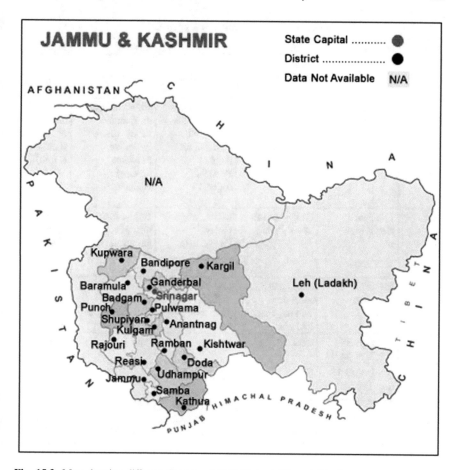

Fig. 15.3 Map showing different districts of J&K State and its neighboring areas

June and July are the hottest months of the zone while December, January and February are the coldest. Its sub-zone is outer hills with predominance of brown hill soil, with slightly higher elevation than the subtropical zone.

Mid to High Altitude Intermediate Zone (JK-2)

This zone is subtropical-temperate transition and comprises of the mid and high altitude areas of the Panjal trap. The zone is characterized by monsoon, concentration of precipitation, relatively wetter, cold winters and higher mean annual rainfall than subtropical zone. The soils are mainly spodic. It encompasses all the areas above outer hills, including the districts of Doda, Poonch, parts of Rajouri, Udhampur and Kathua. The zone varies in elevation from 800 to 1500 m amsl in mid altitude and up to 4000 m amsl in higher altitude. River Chenab and its tributaries constitute the major drainage base. However, upper parts of Kathua

Table 15.3 Major characteristics of agro-climatic zones of J&K

S. no.	Particulars	Jammu		Kashmir	Ladakh
		JK-1 (sub-tropical)	JK-2 (Intermediate)	JK-3 (Temperate)	JK-4 (Cold arid)
1.	Geographical distribution	Jammu district, lower parts of Udhampur, Rajouri, Kathua, Poonch districts	District Doda, all outer hills of Jammu Division and parts of Poonch, Rajouri	All six districts of Kashmir valley viz., Anantnag, Pulwama, Srikangar, Budgam	Two districts of Ladakh (Leh, Kargil)
2.	Principal crops/fruits	Paddy, maize, wheat, oats	Maize, wheat, barley, Paddy, oats, oilseeds	Paddy, maize, oilseeds, temperate fruits almond, saffron	Barley, wheat, alfalfa, apricot
3.	Major livestock	Cross and local cow, buffalo, sheep and goat	Local cow, buffalo, crossbred cow	Crossbred and local cow, sheep and goat	Local and crossbred cow, yak, pashmina
4.	Average land holdings (ha)	0.99	0.93	0.53	1.08
5.	Net irrigated area (%)	36	10	62	100
6.	Major rivers	Ravi, Tawi	Chenab	Jhelum	Indus, Shyok
7.	Altitude (m amsl range)	300–1350	800–1500	2400–3000	3500–8400
8.	Average annual rainfall	1069	1649	789	83
9.	Temperature (°C)				
	Minimum	32.1	31.4	24.5	17.4
	Maximum	13.6	11.5	1.2	−7.0
10.	Thermal index	Mild	Mild	Cold	Very cold
11.	Hydric index	Humid	Humid	Humid	Arid

district drain into Ravi. Its sub-zone marks the limit between valley temperate and cold arid zone. The intermediate zone marks almost the last line of South Western monsoon in summer and similarly the last line of North Western disturbance in winter. In summer, the zone, therefore, receives more rainfall than subtropical and valley temperate zone.

Kashmir Region (Mid to High Altitude Temperate Zone) (JK-3)

Kashmir region or temperate zone essentially covers the valley of Kashmir comprising of the districts Anantnag, Pulwama, Srinagar, Budgam, Baramulla and Kupwara. This zone experiences wet and often severe winters with frost, snow and rain and relatively dry and warm summer. Snowfall, an important form of precipitation, helps to maintain adequate moisture supply during summer when rainfall is scanty. The valley temperate zone encompasses the areas of varied relief. The plain valleys have an altitude of 1560 m amsl, which rises to 1950 m in low altitude Karewas in mid belts, 2400–3000 m in the upper belts and to 4200 m in snow bound areas. The soils of Kashmir valley are alluvial in nature with irrigated area of about 62%. The salient meteorological features of temperate zone show that the zone receives annual rainfall of around 680 mm, of which nearly 70% is received in winter and spring seasons (from December to May). The overall average temperature in different months varies from 1.2 to 24.5 °C with cold thermal index and humid hydric index.

Ladakh Region (Cold Arid Zone) (JK4)

In India, arid zone comprises 3,87,390 km^2 area of which 1,07,545 km^2 lie in the cold arid region of Western Himalayas. The rest of area is hot arid of Indo-Gangetic plains and peninsular India (Directorate of Economics and Statistics, J&K,

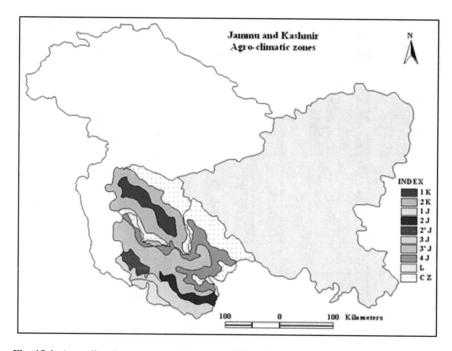

Fig. 15.4 Agro-climatic zone map of Jammu and Kashmir

Table 15.4 Characteristics of micro agro-climatic zones of Jammu and Kashmir

Zone code	Altitude (in m)	Dominant soil group	Crops Grown	Area (in km^2)	Productivity (Q/ha)			Average Precipitation (in mm)	Temperature (°C)	
					Rice	Maize	Wheat		M. Max	M. Min
1K	1000–1700	68, 81, 70	Rice, maize, mustard	4443.2	21.87	12.9	12.37	660	19.34	7.32
2K	1700–3000	58, 81, 18	Rice, maize, orchards	6045.21	20.58	14.63	14.01	967	16.62	5.92
3K	Above 3000	55, 18, 36	–	1000.94	–	–	–	1476	13.12	−0.23
3'K	Above 3000	17, 87, 58	–	4168.15	–	–	–	914	14.31	0.14
1J	Below 500	125, 124, 130	Basmati rice, wheat	1336.49	25.76	21.41	18.91	1386	28.09	17.39
2J	500–1000	115, 116, 86	Maize, wheat, rice	1570.67	21.03	18.97	18.87	1336	27.23	16.90
2'J	500–1000	116, 86, 115	Maize, rice, wheat	1166.68	21.12	17.03	19.09	1210	27.80	17.50
3J	1000–2000	116, 115, 125	Wheat, rice, maize	4131.09	22.99	14.9	18.42	1412	30.19	17.27
3'J	1000–1700	115, 86, 116	Maize, rice, wheat	7432.03	20.38	12.35	19.1	1592	25.48	14.84
4J	1700–3000	17, 116, 115	Maize, wheat, rice	5442.42	20.05	15.61	18.84	1387	22.32	9.43
5J	Above 3000	86, 17, 48	–	376.01	–	–	–	976	19.60	9.52

(continued)

Table 15.4 (continued)

Zone code	Altitude (in m)	Dominant soil group	Crops Grown	Area (in km²)	Productivity (Q/ha)			Average Precipitation (in mm)	Temperature (°C)	
					Rice	Maize	Wheat		M. Max	M. Min
5'J	Above 3000	17, 115, 116	–	5843.8	–	–	–	1642	26.60	14.73
5"J	Above 3000	GG, 55, 17	–	440.91	–	–	–	1329	20.20	7.09
L	Above 3000	2, GG	Millets, barley, wheat	93,531	–	–	19.39	157	11.11	–2.53

2010-11). The cold arid region of Western Himalaya mainly comprises Ladakh area of Jammu and Kashmir State and some parts of Lahul-Spiti sub-division in Himachal Pradesh. The region in J&K lies in the northern most tip of Asian sub-continent between Karakoram and greater Himalayan ranges and is interwoven with nude and rugged mountains. The major characteristics of agro-climatic zones of J&K are summarized in Table 15.3.

Micro Agro-climatic Zones of Jammu and Kashmir

Each province of the state of Jammu and Kashmir has been further delineated into micro agro-climatic zones (Ganai et al. 2014). The agro-climatic zone map of Jammu and Kashmir is shown in Fig. (15.4). The characteristics of these zones are tabulated in Table 15.4 "K, J and L" represent Kashmir, Jammu and Ladakh region.

Chapter 16
Synoptic Meteorology

Abstract Weather observations, taken on the ground or on ships, and in the upper atmosphere with the help of balloon soundings, represent the state of the atmosphere at a given time. When the data are plotted on a weather map, a synoptic view of the weather is obtained. Hence day-to-day analysis and forecasting of weather is known as synoptic meteorology. The chapter explains synoptic meteorology, synoptic charts and synoptic weather systems in India.

Keywords Synoptic meteorology · Synoptic charts · Concept

16.1 Introduction

The word "Synoptic" means "observe together" or "observe at a common point". Weather observations, taken on the ground or on ships, and in the upper atmosphere with the help of balloon soundings, represent the state of the atmosphere at a given time. When the data are plotted on a weather map, a synoptic view of the weather is obtained. Hence day-to-day analysis and forecasting of weather is known as synoptic meteorology.

16.2 Synoptic Chart

A synoptic chart is a map which summarizes atmospheric conditions (temperature, precipitation, wind speed and direction, atmospheric pressure and cloud cover) over a wide area at particular time locations. Synoptic charts are used for observation of movement of weather formations and prediction their future behavior and movements. Charts are updated at least every 6 h, and this allows meteorologists to make more accurate predictions. The symbols used for the preparation of synoptic charts are shown in Fig. 16.1.

© Springer International Publishing AG 2017
L. Ahmad et al., *Experimental Agrometeorology: A Practical Manual*,
https://doi.org/10.1007/978-3-319-69185-5_16

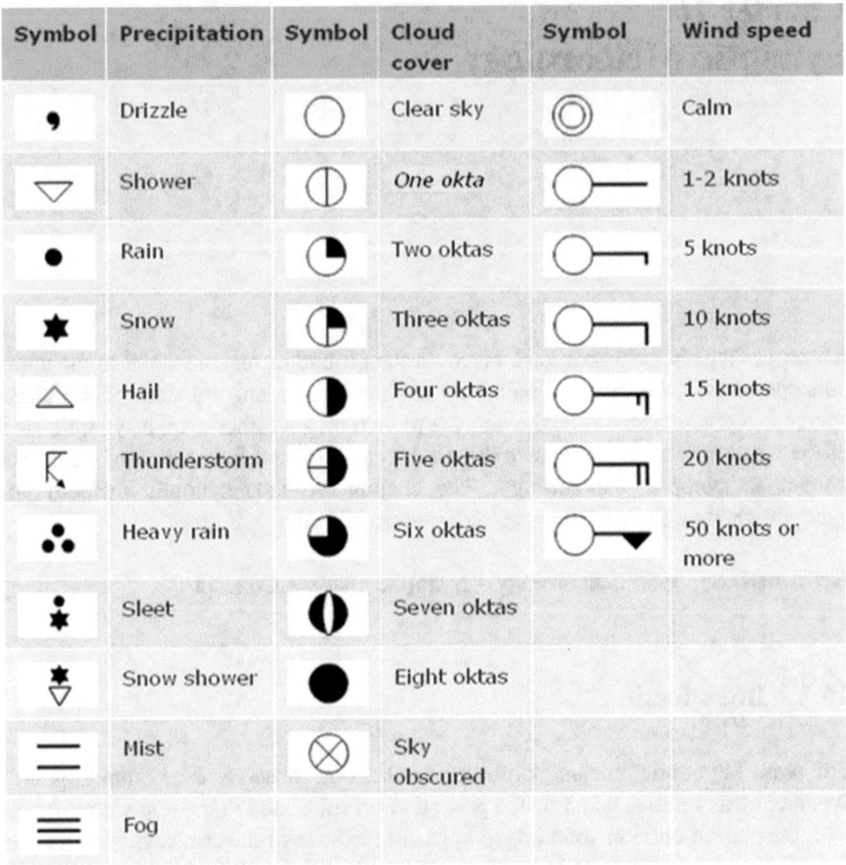

Symbol	Precipitation	Symbol	Cloud cover	Symbol	Wind speed
🖊	Drizzle	◯	Clear sky	◎	Calm
▽	Shower	◐	One okta	◯—	1-2 knots
●	Rain	◕	Two oktas	◯┑	5 knots
★	Snow	◕	Three oktas	◯─┐	10 knots
△	Hail	◑	Four oktas	◯─┐	15 knots
⫧	Thunderstorm	◕	Five oktas	◯─╖	20 knots
⦂	Heavy rain	◕	Six oktas	◯─▼	50 knots or more
⦂	Sleet	◖	Seven oktas		
⛇	Snow shower	●	Eight oktas		
—	Mist	⊗	Sky obscured		
≡	Fog				

Fig. 16.1 Symbols used for the preparation of synoptic charts

16.3   Synoptic Weather Systems in India

The climate of India consists of our seasons, two major seasons and two transitional seasons:

- South-west Monsoon (June–September),
- Transition-I, post-monsoon season (October–November),
- Winter season (December–February),
- Transition-II, Pre-monsoon season (March–May).

South-west Monsoon

The south-west summer monsoons (popularly known as monsoon) occur from June to September. It sets in over Kerala and then advances from south to north and from east to west. Normal onset date is 1″ June. The south-west monsoon is generally expected to begin around the start of June and fade down by the end of September. Nearly 75% of the annual rainfall in India is due to south-west monsoon. Apart from south-eastern part of the Indian peninsula and Jammu and Kashmir, south-west monsoon is the principal source of rain in India. The monsoon accounts for 80% of the rainfall in India. Indian agriculture is heavily dependent on the rains, for growing crops especially like cotton, rice, oilseeds and coarse grains.

Post-monsoon season or North east monsoon season

Major weather disturbances in this season includes NE monsoon, cyclonic storms over Indian seas and low level easterly waves at southern latitudes. The NE monsoon causes rainfall over peninsular India and parts of Sri Lanka. Cities, which get less rain from the South-west Monsoon. About 50–60% of the rain received by the state of Tamil Nadu is from the Northeast Monsoon. In Southern Asia, the northeastern monsoons take place from December to early March.

Winter season

Major Semi permanent synoptic feature is the sub-tropical westerly jet stream and the subtropical ridge. Both of them are seen over Indian longitudes. Major Synoptic weather systems in this season are Western disturbance, mid and upper tropospheric westerly waves at the northern latitudes and sometimes easterly waves at southern latitudes. During winter months, the Western Himalayas and adjoining northern parts of India experience cloudiness and rainfall/snowfall in association with weather systems commonly known as "Western Disturbances". These are responsible for heavy rainfall and snowfall in the Himalayas and Jammu and Kashmir.

Pre-monsoon Season

There is very little rainfall in India in this season. Some thunderstorms occur mainly in Kerala, west Bengal and Assam. Some cyclonic activity also occurs on the East Coast.

Chapter 17
Agro Meteorological Advisory Service

Abstract The Agro-meteorological Advisory Service (AAS) rendered by India Meteorological Department (IMD), Ministry of Earth Sciences (MoES) is a mechanism to apply relevant meteorological information to help the farmer make the most efficient use of natural resources, with the aim of improving agricultural production; both in quantity and quality. A brief introduction to Agro meteorological Advisory Service, Information Support Systems under AAS and Database provided by AAS is given.

Keywords Advisory services · Information support system · Data base

17.1 Introduction

Weather and climatic information plays a major role before and during the cropping season. The Agro-meteorological Advisory Service (AAS) rendered by India Meteorological Department (IMD), Ministry of Earth Sciences (MoES) is a mechanism to apply relevant meteorological information to help the farmer make the most efficient use of natural resources, with the aim of improving agricultural production; both in quantity and quality. Agro-meteorological advisories have become vital to stabilize yields through management of agro-climatic resources as well as other inputs such as irrigation, fertilizer and pesticides.

The main emphasis of the existing AAS system is to collect and organize climate/weather, soil and crop information, and to combine them with weather forecast to assist farmers in taking management decisions. This has helped to develop and apply operational tools to manage weather related uncertainties through agro-meteorological applications for efficient agriculture in rapidly changing environments.

© Springer International Publishing AG 2017
L. Ahmad et al., *Experimental Agrometeorology: A Practical Manual*,
https://doi.org/10.1007/978-3-319-69185-5_17

17.2 Information Support Systems Under AAS

The information support Systems under AAS include:

- Provision of weather, climate, crop/soil and pest disease data to identify biotic and abiotic stress for on-farm strategies and tactical decisions,
- Provide district specific weather forecast (Rainfall, cloudiness, temperature, Wind speed, Wind direction, relative humidity) up to 5 days with outlook for rainfall for remaining two days of a week,
- Translate weather and climate information into farm advisories using existing research knowledge on making more efficient use or Climate and soil resources through applications of medium range weather forecast to maximize benefits of benevolent weather conditions and alleviate the adverse impacts of malevolent weather events. A broad spectrum of advisories includes weather sensitive farm operations such as sowing, transplanting of crops, fertilizer application based on wind condition and intensity of rain, pest and disease control, intercultural quantum and timing of irrigation using meteorological threshold and advisories for timely harvest of crops,
- Introduction of technologies such as crop simulation model based decision support system for agro-meteorologists to adapt agricultural production systems to changing weather and climate variability and to the increasing scarcity of input such as water, seed, fertilizer, pesticide etc.,
- Develop effective mechanism for on time dissemination of agro-met advisories to farmers,
- Effective training, education and extension on all aspects of agricultural meteorology.

17.3 Database Provided by AAS

Preparation of data base of the following can help identify the cropping pattern, production potential and possible risks for a particular cropping pattern of an area:

- Agro-climatic Classification
- Averages and extremes of different weather parameters
- Heat Waves/Cold Waves
- Heavy Rainfall/Snowfall
- Frequency and Severity of other disastrous weather phenomena (Dust-storms, Thunderstorms, Hail Storms, Squalls etc.)
- Dry and Wet Spells
- Assured Rainfall
- Evaporation Potential
- Evapotranspiration
- Soil Temperature

- Sowing Dates
- Percentage of Soil Moisture
- Grass Minimum Temperature
- Hours of Bright Sunshine
- Global Solar Radiation
- Wind Roses.

The agrometeorological services can provide these data base (for the stations under their jurisdiction) to the planners in the beginning of the crop seasons to help them in formulating strategies for minimizing the adverse effects of an unfavorable weather pattern.

Chapter 18
Crop Yield Forecast Models

Abstract In agro-meteorological research, the crop models basically help in testing scientific hypothesis, highlight where information is missing, organizing data and integrating across disciplines. The crop growth models can be used to predict crop performance in regions where the crop has not been grown before or not grown under optimal conditions. The chapter introduces Crop Simulation Models and their applications.

Keywords Crop yield forecast · Simulation models · Input data
Application

18.1 Introduction

Crop growth and development depend on a large number of biotic and abiotic factors. The yield of any crop can be forecast from the general growth pattern of the crop and effect of biotic and abiotic factors if long term historic and temporal data are available with their effects on each component of growth and development.

Prediction of yield has its short term and long term benefits. The short-term benefits are

1. The likely yield, and whether that would be sufficient to meet the demand
2. If the yield is not sufficient, what other sources are to be tapped for meeting the requirement.

The long term benefits of yield forecast lie in export-import implication and strategies to meet the adverse conditions through long range crop planning.

L. Ahmad et al., *Experimental Agrometeorology: A Practical Manual*,
https://doi.org/10.1007/978-3-319-69185-5_18

18.2   Crop Simulation Models

Crop Simulation Models (CSM) are computerized representations of crop growth, development and yield, simulated through mathematical equations as functions of soil conditions, weather and management practices (Hoogenboom et al. 2004).

In agro-meteorological research, the crop models basically help in testing scientific hypothesis, highlight where information is missing, organizing data and integrating across disciplines. The crop growth models can be used to predict crop performance in regions where the crop has not been grown before or not grown under optimal conditions. Such applications are of value for regional development and agricultural planning in developing countries. The crop Simulation Models are usually climatological models or water stress models or crop growth model. The climatological models are related to location specific climate and soil factors. The moisture stress models may be either Single stage model or Multistage model-additive or multiplicative. In single stage model the period from sowing/planting to harvest is considered as one stage. In the multistage models, the growth period is divided into specific stages (phenophases).

18.3   Input Data Requirement

For running a crop simulation model, the necessary data required includes weather data, crop data, soil data of the location where the crop is sown and the data on management practices adopted from time to time during the crop growth period. The input data required for crop simulation is presented in Fig. 22.1.

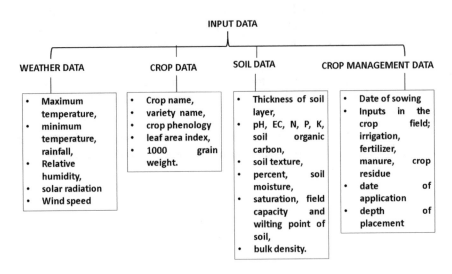

18.4 Possible Applications of Crop Model

- Estimation of potential yields;
- Estimation of yield gaps: principal causes and their contribution;
- Yield forecasting;
- Impact assessment of climatic variability and climatic change;
- Optimizing management—Dates of planting, variety, irrigation and nitrogen fertilizer;
- Environmental impact—percolation, N losses, GHG emissions, SOC dynamics;
- Plant type design and evaluation;
- Genotype by environment interactions.

18.5 Examples of Crop Models

Some of the commonly used crop simulation models are given below:

Model name	Full form
DSSAT	Decision Support System for Agro technology Transfer
IBSNAT	International Benchmark Sites Network for Agrotechnology Transfer
EPIC	Environmental Policy Integrated Climate Model
CROPSYST	Cropping Systems Simulation Model
WOFOST	World Food Studies
ADEL	Architectural model of Development (based on L-System) for graminae

Chapter 19
Measurement of Soil Moisture

Abstract The water present in the interstices of the soil matrix constitutes the soil moisture. It is one of the most important constituents of the soil. The water stored in the root zone is used by the plants for photosynthesis, evapotranspiration and building of its tissues. Soil properties such as soil moisture content, texture, structure density, fertility, salinity aeration, drainage and temperature have a pronounced effect on growth and yield responses of a crop. The chapter describes the methods and techniques for determination of soil moisture content and expression of soil moisture content.

Keywords Soil moisture · Measurement · Types · Layout

19.1 Introduction

The water present in the interstices of the soil matrix constitutes the soil moisture. It is one of the most important constituents of the soil. The water stored in the root zone is used by the plants for photosynthesis, evapotranspiration and building of its tissues. Soil properties such as soil moisture content, texture, structure density, fertility, salinity aeration, drainage and temperature have a pronounced effect on growth and yield responses of a crop.

19.2 Terminology

Saturation Capacity

Saturation capacity is the maximum water holding capacity of the soil when all the pores are saturated with water.

© Springer International Publishing AG 2017
L. Ahmad et al., *Experimental Agrometeorology: A Practical Manual*,
https://doi.org/10.1007/978-3-319-69185-5_19

Field Capacity

The moisture content in the soil after drainage of gravitational water is called field capacity.

Permanent Wilting Point

It is the moisture content at which the plants cannot extract sufficient water from the soil to meet the needs of transpiration and other processes.

Wilting Range

The range of soil moisture content through which plants undergo progressive and irreversible wilting.

Ultimate Wilting Point

The moisture content at which wilting is complete and plants die.

Available Water

Moisture present in the soil between field capacity and permanent wilting point.

19.3   Expression of Soil Moisture Content

Soil moisture content is expressed as weight percent or volume percent. Soil moisture content percent by weight (θ_m) is given as:

$$\theta_m = \frac{\left(Weight\ of\ moist\ soil\ sample - Weight\ of\ ovendry\ sample\right)}{Weight\ of\ oven\ dry\ sample} \times 100$$

Soil moisture content percent by volume (θ_v) is then calculated as:

$$\theta_v = \theta_m \times Soil\ Bulk\ density$$

19.4   Determination of Soil Moisture Content

Soil moisture measurements are performed on a regular basis in agro-meteorological observatories. These are necessary for irrigation scheduling and soil water management.

Layout for Moisture Observation

The soil moisture content is determined at a weekly interval throughout the year. The depths for measurement are taken as 7.5, 15, 30, 45 and 60 cm. A suitable plot size of 300 cm × 200 cm is marked out in observations. The plot should be kept free from vegetation. The soil samples should be collected systematically about 30 cm away from previous sampling point. The layout for moisture content measurements is given in Fig. 19.1.

Soil moisture observations are recorded from the field where the major crop of the region is grown. The soil samples are collected once a week from the time of sowing/planting to just after harvest. In case of irrigated field besides weekly, the observations are also taken at all depths just before and two days after the irrigation or rainfall.

Methods of Soil Moisture Content Determination

1. **Gravimetric Method**

In this method, soils samples of known weight or volume are used. A soil auger or sampling tube is used for the collection of samples. The samples are then placed in air tight aluminum containers and then dried in an oven at 105 °C for 24 h. The samples are then cooled to room temperature and weighed again. The difference in weight is the moisture content by weight.

2. **Tensiometer Method**

The tensiometer is an instrument that to provides a continuous indication of the soil-moisture tension. Soil moisture tension is the tenacity with which water is held in soils. The Tensiometer consists of a porous ceramic cup filled with water, connected through a tube to a vacuum gauge or a manometer and at the. The top of this cup has an air-tight removable seal (Fig. 19.2). The cup is placed in the soil where the suction measurement is to be made. The water inside the cup on coming into contact with the soil water tends to equilibrate with it through the pores in the ceramic walls.

Initially, the water contained in the tensiometer is generally at atmospheric pressure (essentially, 0 bars tension). Soil water on the other hand is generally at sub-atmospheric pressure (or higher tension). Thus, it exercises a suction, which draws out a certain amount of water from the rigid and air tight tensiometer. As a

200 cm

1	6	11	16	21	26	31	36	41	46	51
2	7	12	17	22	27	32	37	42	47	52
3	8	13	18	23	28	33	38	43	48	
4	9	14	19	24	29	34	39	44	49	
5	10	15	20	25	30	35	40	45	50	

300 cm

Fig. 19.1 Layout of the moisture observation

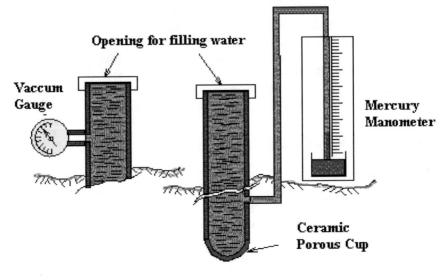

Fig. 19.2 Vacuum gauge and manometer-type tensiometers

result, the pressure inside the tensiometer falls below atmospheric pressure. The sub-pressure is indicated by a vacuum gauge or manometer. The changes are reflected in the tensiometer reading assoil moisture is depleted by plant uptake or as it is replenished by rainfall or irrigation.

Fig. 19.3 Neutron moisture meter probes at different depths

3. Neutron Moisture Meter Method

It is used for the in situ moisture measurement of the soil. It is based on the principle that the number of hydrogen nuclei present in a unit volume of soil is proportional to the number of water molecules present in the same volume. The Neutron Moisture Meter has two principal parts—a probe and a scaler. The probe is lowered in the soil at suitable depths (Fig. 19.3). The probe is a source fast neutrons and a detector of slow neutrons. The scaler is battery equipped and is used to monitor slow moving neutrons, which is proportional to soil water content. A mixture of radium and beryllium or americium and beryllium is used as a source of fast moving neutrons.

Agro-Meteorological Glossary

Terms	Definition
Types of forecast	
Now casting	A short-range forecast having a lead time/validity of less than 24 hrs
Short range forecasts	Forecasts having a lead time/validity period of 1–3 days
Medium range forecasts	Forecasts having a lead time/validity period of 4–10 days
Long range /Extended range forecasts	Forecasts having a lead time/validity period beyond 10 days Usually this is being issued for a season
Seasons	Meteorological seasons over India are: Winter Season: January–February
	Pre-Monsoon Season: March–May
	Southwest Monsoon Season: June–September
	Post Monsoon Season: October–December
Monsoon	
Monsoon	The seasonal reversal of winds and the associated rainfall
Southwest monsoon	The southwesterly wind flow occurring over most parts of India and Indian Seas gives rise to southwest monsoon over India from June to September
Onset of southwest monsoon	Commencement of rainy season with the establishment of monsoon flow pattern
	Normal date for Onset of southwest monsoon
	South Andaman Sea: 20 May
	Kerala: 1 June
	Mumbai: 10 June
	New Delhi: 29 June
	Entire country: 15 July
Withdrawal of southwest monsoon	Cessation of southwest monsoon rainfall
	Normal date of withdrawal from extreme west Rajasthan is 15 September

(continued)

© Springer International Publishing AG 2017
L. Ahmad et al., *Experimental Agrometeorology: A Practical Manual*,
https://doi.org/10.1007/978-3-319-69185-5

(continued)

Terms	Definition
Northeast monsoon	With the withdrawal of the southwest monsoon from the northern and central India and the northern parts of the Peninsula by the first half of the October, the wind pattern rapidly changes from southwesterly to northeasterly and hence the term "Northeast Monsoon" is used to describe the period October–December
Weekly/seasonal rainfall distribution	
Excess	Percentage departure of realized rainfall from normal rainfall is +20% or more
Normal	Percentage departure of realized rainfall from normal rainfall is between −19 and +19%
Deficient	Percentage departure of realized rainfall from normal rainfall is between −20 and −59%
Scanty	Percentage departure of realized rainfall from normal rainfall is between −60 and −99%
No rain	Percentage departure of realized rainfall from normal rainfall is −100%
Rainfall distribution on All India Scale	
Normal	Percentage departure of realized rainfall is within ±10% of the Long Period Average
Below normal	Percentage departure of realized rainfall is <10% of the Long Period Average
Above normal	Percentage departure of realized rainfall is >10% of the Long Period Average
All India drought year	When the rainfall deficiency is more than 10% and when 20–40% of the country is under drought conditions, then the year is termed as All India Drought Year
All India severe drought year	When the rainfall deficiency is more than 10% and when the spatial coverage of drought is more than 40% it is called as All India Severe Drought Year
Temperature	
Temperature	The temperature of a body is the condition which determines its ability to communicate heat to other bodies or to receive heat from them
Air temperature	The temperature measured in an enclosed space allowing free flow of air and not directly exposed to sunlight where the thermometer is kept at a height of 1.2 m above the surface
Maximum temperature	The highest air temperature recorded in a day
Minimum temperature	The lowest air temperature recorded in a day
Dew point temperature	The temperature to which moist air must be cooled, during a process in which pressure and moisture content of the atmosphere remain constant
Freezing point	The constant temperature in which the solid and liquid forms of pure water are in equilibrium at Standard Atmospheric Pressure

(continued)

(continued)

Terms	Definition
Normal	Departure of minimum/maximum temperature from normal is +1 to −1 °C
Above normal	Departure of minimum/maximum temperature from normal is +2 °C
Appreciably above normal	Departure of minimum/maximum temperature from normal is +3 to +4 °C. The normal maximum temperature should be 40 °C or less
Markedly above normal	Departure of minimum/ maximum temperature from normal is from +5 to +6 °C. The normal maximum temperature should be 40 °CC or less
Hot day	Whenever the maximum temperature remains 40 °C or more and minimum remains 5 °C or more above normal, provided, it is not satisfying the heat wave criteria
Heat wave	Departure of maximum temperature from normal is +4 °C to +5 °C or more for the regions where the normal maximum temperature is more than 40 °C and departure of maximum temperature from normal is +5 to +6 °C for regions where the normal maximum temperature is 40 °C or less
	(Heat Wave is declared only when the maximum temperature of a station reaches at least 400 C for plains and at least 300 C for Hilly regions)
	When actual maximum temperature remains 45 °C or more irrespective of normal maximum temperature, heat wave is declared
Severe heat wave conditions	Departure of maximum temperature from normal is +6 °C or more for the regions were the normal maximum temperature is more than 40 °C and +7 °C or more for regions were the normal maximum temperature is 40 °C or less
	(Heat Wave is declared only when the maximum temperature of a station reaches at least 400 C for plains and at least 300 C for Hilly regions)
Cold day	In the plains of north India, foggy conditions prevail during winter for several days or weeks. The minimum temperature on these days remains above normal, while maximum temperature remains much below normal. This creates cold conditions for prolonged period
	When maximum temperature is less than or equal to 16 °C in Plains, it will be declared "Cold Day"
Cold wave	Wind chill factor is taken into account while declaring the cold wave situation
	The wind chill effective minimum temperature (WCTn) is defined as the effective minimum temperature due to wind flow. For ex. When the minimum temperature is 15 °C and the wind speed is 10 mph, WCTn will be 10.5 °C
	Departure of WCTn from normal minimum temperature is from −5 to −6 °C where normal minimum temperature >10 °C

(continued)

(continued)

Terms	Definition
	and from −4 to −5 °C elsewhere, Cold Wave is declared. For declaring cold wave etc. WCTn only is used and when it is <10 °C only, cold wave is considered (this criterion does not hold for coastal stations)
	Also, cold wave is declared when WCTn is <0 °C irrespective of the normal minimum temperature for those stations.
Severe cold wave	Departure of WCTn from normal minimum temperature is −7 °C or less for the regions where normal minimum temperature is >10 and −6 °C or less elsewhere. (departure of WCTn from normal minimum temperature is from −5 to −6 °C where normal minimum temperature >10 °C and from −4 to −5 °C elsewhere)
Precipitation	Precipitation whether it is rain or snow is expressed as the depth to which it would cover a horizontal projection of the earth's surface, if there is no loss by evaporation, run–off or infiltration and if any part of the precipitation falling as snow or ice were melted. It is expressed in the units of mm or cm
Rainfall	Liquid rainfall is expressed as the depth to which it would cover a horizontal projection of the earth's surface, if there is no loss by evaporation, run–off or infiltration. It is expressed in terms of mm or cm
Snowfall	Snowfall is measured either as the depth of snow which has fallen in a stated period, or melted and measured as water
Relative humidity	Relative Humidity is the ratio of the actual quantity of moisture at a certain temperature and pressure to the maximum it can hold at the same temperature and pressure
Spatial distribution of rainfall	
Widespread (most places)	75% or more number of stations of a region (usually a meteorological sub-division) reporting at least 2.5 mm rainfall
Fairly widespread (many places)	51–74% number of stations of a region (usually a meteorological sub-division); reporting at least 2.5 mm rainfall
Scattered (at a few places)	26–50% number of stations of a region (usually a meteorological sub-division) reporting at least 2.5 mm rainfall
Isolated (at isolated places)	25% or less number of stations of a region (usually a meteorological sub-division) reporting at least 2.5 mm rainfall
Mainly dry	No station of a region reported rainfall
Intensity of rainfall	
No rain	Rainfall amount realized in a day is 0.0 mm
Trace	Rainfall amount realized in a day is between 0.01 and 0.04 mm
Very light rain	Rainfall amount realized in a day is between 0.1 and 2.4 mm
Light rain	Rainfall amount realized in a day is between 2.5 and 7.5 mm
Moderate rain	Rainfall amount realized in a day is between 7.6 and 35.5 mm

(continued)

(continued)

Terms	Definition
Rather heavy	Rainfall amount realized in a day is between 35.6 and 64.4 mm
Heavy rain	Rainfall amount realized in a day is between 64.5 and 124.4 mm
Very heavy rain	Rainfall amount realized in a day is between 124.5 and 244.4 mm
Extremely heavy rain	Rainfall amount realized in a day is more than or equal to 244.5 mm
Exceptionally heavy rainfall	This term is used when the amount realized in a day is a value near about the highest recorded rainfall at or near the station for the month or season. However, this term will be used only when the actual rainfall amount exceeds 12 cm
Rainy Day	Rainfall amount realized in a day is 2.5 mm or more
Synoptic systems	
Cyclonic circulation (Cycir)	Atmospheric wind flow in upper levels associated with any low-pressure system. The wind flow is counterclockwise in the Northern Hemisphere and clockwise in the Southern Hemisphere
Anticyclonic circulation	Atmospheric wind flow in upper levels associated with any high-pressure system. The wind flow is clockwise in the Northern Hemisphere and counterclockwise in the Southern Hemisphere
Low pressure Area (lopar)/ well marked lopar	Area in the atmosphere in which the pressures are lower than those of the surrounding region at the same
Depression	Intense low pressure system represented on a synoptic chart by two or three closed isobars at 2 hPa interval and wind speed from 17 to 27 Kts at sea and two closed isobars in the radius of 3° from the center over land
Deep depression	Intense low pressure system represented on a synoptic chart by two or three closed isobars at 2 hPa interval and wind speed from 28 to 33 Kts at sea and three to four closed isobars in the radius of 3° from the center over land
Cyclonic storm	Intense low pressure system represented on a synoptic chart by more than four closed isobars at 2 hPa interval and in which the wind speed on surface level is in between 34 and 47 Kts
Severe cyclonic strom	Intense low pressure system represented on a synoptic chart by more than four closed isobars at 2 hPa interval and in which the wind speed on surface level is in between 48 and 63 Kts
Super cyclonic storm	Very Severe Cyclonic Storm Intense low pressure system represented on a synoptic chart by more than four closed isobars at 2 hPa interval and in which the wind speed on surface level is in between 64 and 119 Kts
	Intense low pressure system represented on a synoptic chart by more than four closed isobars at 2 hPa interval and in which the wind speed on surface level is 120 Kts and above

(continued)

(continued)

Terms	Definition
Western disturbance	Weather disturbances noticed as cyclonic circulation/trough in the mid and lower tropospheric levels or as a low-pressure area on the surface, which occur in middle latitude westerlies and originate over the Mediterranean Sea, Caspian Sea and Black Sea and move eastwards across north India
Western depression	Weather system which originate over the Mediterranean Sea, Caspian Sea and Black Sea and approach northwest India and is defined by two or more closed isobars on the surfac
Induced low	Under the influence of the western disturbance, sometimes a low is developed to the south of the system called as induced low
Induced cyclonic circulation	Under the influence of the western disturbance, sometimes a cyclonic circulation is developed to the south of the system called as induced cyclonic circulation
Trough	A line or curve along which the atmospheric pressure is minimum. Pressure increases on both sides of the line or curve
Trough in westerlies	A moving wave perturbation in mid latitude regions which are present throughout the year which move from west to east and entire globe. These systems generally affect the northern parts of India
Trough in easterlies	A moving wave perturbation in the equatorial easterly wave, moving from east to west
Easterly waves	A shallow trough disturbance in the easterly current of the tropics, more in evidence in the upper level winds than in surface pressure, whose passage westwards is followed by a marked intensification of cloudy, showery weather. The southern peninsular region is affected by easterly waves
Shear line	A line or narrow zone across which there is an abrupt change in the horizontal wind component; a line of maximum horizontal wind shear
Ridge	An elongated area of relatively high atmospheric pressure, almost always associated with and most clearly identified as an area of maximum anticyclonic curvature of wind flow
High/High pressure area	Area in the atmosphere in which the pressures are higher than those of the surrounding region at the same level and is represented on a synoptic chart by a system of, at least, one closed isobar
Wind-discontinuity	A line across which there is an abrupt change in wind direction
Troposphere	An atmospheric layer in which all significant weather phenomena occur. The troposphere is characterized by decreasing temperature with height
Sky conditions	Reported in terms of Octa wherein the sky is divided into 8 equal parts
0 octa	Clear sky
1–2 octa of sky covered	Mainly clear

(continued)

(continued)

Terms	Definition
3–4 octa of sky covered	Partly cloudy
5–7 octa of sky covered	Generally cloudy
>7 octa of sky covered	Cloudy
Winds	
Wind	Atmospheric motion characterized by direction and speed. The direction of the wind is the direction from which the wind approaches the station (Example Northerly wind—Wind approaching the station from North)
Gales	A gale is a very strong wind (34–47 knots)
Squall	A sudden increase of wind speed by atleast 3 stages on the Beaufort Scale, the speed rising to force 6 or more, and lasting for atleast one minute
Gust	A rapid increase in the strength of the wind relative to the mean strength at the time
Weather phenomena	
One or two spells of rain	In a 24 h time, rainfall occurring with a frequency of 1–2 spells.
A few spells of rain	In a 24 h time, rainfall occurring with a frequency of more than 2 spells but with well-defined dry spells in between
Intermittent rain	In a 24 h time, rainfall occurring with a frequency more than that defined in "A Few Spells" but is discontinuous and without presenting the character of a shower
Drizzle	Liquid precipitation in the form of water drops of very small size (by convention, with radius of water drops between about 100 and 500 μm)
Rain	Liquid precipitation in the form of water drops of radius between about 500 and 2500 μm
Shower	Solid or liquid precipitation from a vertically developed cloud is designated a shower and is distinguished from the precipitation, intermittent or continuous, from layer clouds. Showers are often characterized by short duration and rapid fluctuations of intensity (by convention, with radius of water drops more than 2500 μm)
Hail	Solid precipitation in the form of balls or pieces of ice (hailstones) with diameters ranging from 5 to 50 mm or even more
Thunderstorm	One or more sudden electrical discharges manifested by a flash of light (Lightning) and a sharp rumbling sound (thunder)
Dust-storm	An ensemble of particles of dust or sand energetically lifted to great heights by a strong and turbulent wind
	Surface visibility is reduced to low limits; the qualification for a synoptic report is visibility below 1000 m
Mist	Mist is a phenomenon of small droplets suspended in air

(continued)

(continued)

Terms	Definition
Dew	Condensation of water vapor on a surface whose temperature is reduced by radiational cooling to below the DEW-POINT of the air in contact with it
Fog	Fog is a phenomenon of small droplets suspended in air and the visibility is one kilometer or less
Frost	Frost occurs when the temperature of the air in contact with the ground, or at thermometer-screen level, is below the freezing-point of water ('ground frost' or 'air frost', respectively). The term is also used of the icy deposits which may form on the ground and on objects in such temperature conditions (glaze, hoar-frost)
Haze	Haze is traditionally an atmospheric phenomenon where dust, smoke and other dry particles obscure the clarity of the sky
Smog	Smoke and fog together reduce the visibility
Squally weather	Squally weather is meant to cover occasional or frequent squalls with rain or persistent type of strong gusty winds (mean wind speed not less than 20 knot) accompanied by rain. Such conditions are associated with low pressure systems or onset and strengthening of monsoon
Visibility	The greatest distance that prominent objects can be seen and identified by unaided, normal eyes
Part of the day	
Early hours of (date)	0000–0400 h IST
Morning	0400–0800 h IST
Forenoon	0800–1200 h IST
Afternoon	1200–1600 h IST
Evening	1600–2000 h IST
Night	2000–2400 h IST
Some common terms	
Condensation	The process of formation of a liquid from its vapors; in meteorology, the formation of liquid water from water vapor
Convection	A mode of heat transfer within a fluid, involving the movement of substantial volumes of the substance concerned
Freezing level	Commonly the lowest height above mean sea level at which, for a given place and time, the air temperature is 0 °C
Isobar	A line of constant (atmospheric) pressure
Isohyet	A line of constant rainfall amount
Isotach	A line of constant wind speed
Isogon	A line of constant wind direction
Isotherm	A line of constant temperature
Streamline	A line which is parallel to instantaneous direction of the wind vector at all points along it
Jet stream	A fast narrow current of air, generally near the tropopause, characterized by strong vertical and lateral wind shears

(continued)

(continued)

Terms	Definition
Latent heat	The quantity of heat absorbed or emitted, without change of temperature, during a change of state of unit mass of a material
Radiation	The transmission of energy by electromagnetic waves
Land and sea breezes	Local winds caused by the unequal diurnal heating and cooling of adjacent land and water surfaces; under the influence of solar radiation by day and radiation to the sky at night, a gradient of pressure near the coast is produced. During the day, the land is warmer than the sea and a breeze, the Sea Breeze, blows onshore; at night and in the early morning the land is cooler than the sea and the land breeze blows off shore
Tornado	A tornado is a violent, rotating column of air which is in contact with both the surface of the earth and a cumulonimbus cloud or, in rare cases, the base of a cumulus cloud
Water spout	A waterspout is an intense columnar vortex (usually appearing as a funnel-shaped cloud) that occurs over a body of water and is connected to a cumuliform cloud

Appendix A

Saturation vapor pressures (mm of Hg) in terms of Dry bulb or wet bulb temperature in °C

Air Temp °C	0.0	0.1	0.2	0.3	0.4	0.5	0.6	0.7	0.8	0.9
0	4.6	4.6	4.7	4.7	4.7	4.7	4.8	4.8	4.9	4.9
1	4.9	5.0	5.0	5.0	5.1	5.1	5.1	5.2	5.2	5.3
2	5.3	5.3	5.4	5.4	5.5	5.5	5.5	5.6	5.6	5.6
3	5.7	5.7	5.8	5.8	5.9	5.9	5.9	6.0	6.0	6.1
4	6.1	6.1	6.2	6.2	6.3	6.3	6.4	6.4	6.5	6.6
5	6.5	6.6	6.6	6.7	6.7	6.8	6.8	6.9	6.9	7.0
6	7.0	7.1	7.1	7.2	7.2	7.3	7.3	7.4	7.4	7.5
7	7.5	7.6	7.6	7.7	7.7	7.8	7.8	7.9	7.9	8.0
8	8.1	8.1	8.2	8.2	8.3	8.3	8.4	8.4	8.5	8.5
9	8.6	8.7	8.7	8.8	8.9	8.9	9.0	9.0	9.1	9.1
10	9.2	9.3	9.3	9.4	9.5	9.5	9.6	9.7	9.7	9.8
11	9.8	9.9	10.0	10.0	10.1	10.2	10.2	10.3	10.4	10.5
12	10.5	10.6	10.7	10.7	10.8	10.8	10.9	11.0	11.1	11.2
13	11.2	11.3	11.4	11.5	11.5	11.6	11.7	11.8	11.8	11.9
14	12.0	12.1	12.1	12.2	12.3	12.4	12.5	12.5	12.6	12.7
15	12.8	12.9	13.0	13.0	13.1	13.2	13.3	13.4	13.5	13.5
16	13.6	13.7	13.8	13.9	14.0	14.1	14.2	14.3	14.3	14.4
17	14.5	14.6	14.7	14.8	14.9	15.0	15.1	15.2	15.3	15.4
18	15.5	15.6	15.7	15.8	15.9	16.0	16.1	16.2	16.3	16.4
19	16.5	16.6	16.7	16.8	16.9	17.0	17.1	17.2	17.3	17.4
20	17.5	17.7	17.8	17.9	18.0	18.1	18.2	18.3	18.4	18.5
21	18.7	18.8	18.9	19.0	19.1	19.2	19.3	19.5	19.6	19.7
22	19.8	19.9	20.1	20.2	20.3	20.4	20.6	20.8	20.8	20.9
23	21.1	21.2	21.3	21.5	21.6	21.7	21.9	22.0	22.1	22.2
24	22.4	22.5	22.7	22.8	22.9	23.1	23.2	23.3	23.5	23.6
25	23.8	23.9	24.1	24.2	24.3	24.5	24.6	24.8	24.9	25.1
26	25.2	25.4	25.5	25.7	25.8	26.0	26.1	26.3	26.4	26.6
27	26.7	26.9	27.1	27.2	27.4	27.5	27.7	27.9	28.0	28.2

(continued)

© Springer International Publishing AG 2017
L. Ahmad et al., *Experimental Agrometeorology: A Practical Manual*,
https://doi.org/10.1007/978-3-319-69185-5

(continued)

Air Temp °C	0.0	0.1	0.2	0.3	0.4	0.5	0.6	0.7	0.8	0.9
28	28.4	28.5	28.7	28.9	29.0	29.2	29.3	29.5	29.7	29.9
29	30.1	30.2	30.4	30.6	30.7	30.9	31.1	31.3	31.5	31.7
30	31.8	32.0	32.2	32.4	32.6	32.8	32.9	33.1	33.2	33.5
31	33.7	33.9	34.1	34.3	34.5	34.7	34.9	35.1	35.3	35.5
32	35.7	35.9	36.1	36.3	36.5	36.7	36.9	37.1	37.3	37.5
33	37.7	37.9	38.2	38.4	38.6	38.8	39.0	39.3	39.5	39.7
34	39.9	40.1	40.4	40.6	40.8	41.0	41.3	41.5	41.7	41.9
35	42.2	42.4	42.7	42.9	43.1	43.4	43.6	43.9	44.1	44.3
36	44.6	44.8	45.1	45.3	45.6	45.8	46.1	46.3	46.6	46.8
37	47.1	47.3	47.6	47.9	48.1	48.4	48.6	48.9	49.2	49.4
38	49.7	50.0	50.2	50.5	50.8	51.1	51.3	51.6	51.9	52.2
39	52.5	52.7	53.0	53.3	53.6	53.9	54.2	54.5	54.7	55.0
40	55.3	55.6	55.9	56.2	56.5	56.8	57.1	57.4	57.7	58.1

Dry Bulb temperature and Wet Bulb Depression with Corresponding Relative Humidity

Dry Bulb *c	Depression of wet bulb °C												
	0.5	1.0	1.5	2.0	2.5	3.0	3.5	4.0	4.5	5.0	5.5	6.0	6.5
−1	90	79	69	59	49	39	30	20	10	1			
0	90	81	71	61	52	44	34	25	16	7			
+1	90	81	73	64	55	47	38	29	20	13	4		
2	91	82	73	64	57	49	41	33	24	17	9	1	
3	91	83	74	CC	57	49	43	3G	28	21	14	7	
4	92	83	75	67	59	51	43	35	32	25	18	11	4
5	92	84	76	68	61	53	46	38	31	24	21	15	8
6	92	85	77	70	62	55	48	41	34	27	20	14	12
7	93	85	78	71	64	57	50	44	37	30	24	17	11
a	93	86	79	72	65	59	52	46	39	33	27	21	15
9	93	86	80	73	67	GO	54	48	42	36	30	27	18
10	93	87	81	74	68	62	56	50	44	38	33	30	21
11	9*1	87	81	75	69	63	58	52	¿6	41	36	32	24
12	94	88	82	76	70	65	59	54	48	43	37	36	27
13	94	88	83	77	71	GO	GO	55	50	45	40	37	30
14	94	89	83	78	72	67	62	57	52	47	42	39	32
15	94	89	84	78	73	68	63	58	53	48	42	41	34
16	95	89	84	79	74	69	64	59	55	50	43	43	37
17	95	90	85	eo	75	70	65	61	56	52	47	45	39
18	95	90	85	80	76	71	66	62	57	53	49	46	40
19	90	90	86	81	76	72	67	63	59	54	50	48	42
20	95	91	86	81	77	73	68	64	60	56	52	49	44

(continued)

(continued)

Dry Bulb *c	Depression of wet bulb °C												
	0.5	1.0	1.5	2.0	2.5	3.0	3.5	4.0	4.5	5.0	5.5	6.0	6.5
21	95	91	8G	82	78	73	69	G5	G1	57	53	50	45
22	95	91	87	82	78	74	70	66	62	58	54	52	47
23	96	91	87	83	79	75	71	67	63	59	55	53	48
24	96	91	87	83	79	75	71	68	64	60	57	54	48
25	96	92	88	84	80	76	72	66	65	oi	58	55	51
26	96	92	88	84	80	76	73	69	66	62	59	56	52
27	9G	92	88	84	81	77	73	70	GG	03	59	57	53
28	96	92	88	85	81	77	74	70	67	34	60	58	54
29	96	92	89	85	81	78	74	71	G8	64	61	59	55
30	96	93	89	85	82	78	75	72	68	35	62	61	56
32	96	93	89	86	82	79	7G	73	70	-37	G4	62	58
34	96	93	89	86	83	80	77	74	71	68	65	63	59
36	96	93	90	87	84	81	78	75	72	59	66	64	61
38	96	93	90	87	84	81	78	75	73	70	67	66	62
40	96	94	91	e8	85	82	79	7G	74	71	G9	67	G3
42	97	94	91	88	85	82	80	77	75	72	70	67	65
44	97	94	91	88	86	83	81	78	75	72	70	68	65
46	97	94	91	89	86	83	81	78	76	73	71	69	66
48	97	95	92	89	86	83	81	76	76	74	72	70	67
50	97	95	92	89	87	84	82	79	77	74	72	70	68

Dry Bulb *c	Depression of wet bulb °C												
	7.0	7.5	8.0	8.5	9.0	9.5	10	11	12	13	14	15	16
−1													
0													
+1													
2													
3													
4													
5	2												
6	6												
7	5												
a	9	3											
9	12	7	1										
10	16	10	5										
11	19	14	9	4									
12	22	17	12	7	2								
13	25	20	15	11	G	1							
14	27	23	18	14	9	6	1						
15	30	25	21	17	12	8	4						

(continued)

(continued)

Dry Bulb	Depression of wet bulb °C												
*c	7.0	7.5	8.0	8.5	9.0	9.5	10	11	12	13	14	15	16
16	32	28	24	19	15	11	7						
17	34	30	26	22	19	14	10	3					
18	36	32	26	24	21	17	13	6					
19	38	34	30	27	23	19	16	9	2				
20	40	36	32	29	25	22	18	11	5				
21	42	38	34	31	27	24	20	14	7	1			
22	43	40	36	33	29	26	23	16	10	4			
23	45	41	38	34	31	28	25	18	12	6	1		
24	46	43	39	36	33	30	27	20	15	9	3		
25	47	44	41	38	35	31	28	22	17	11	6	1	
26	49	45	42	39	36	33	33	24	19	13	8	3	
27	50	47	44	41	33	35	32	26	21	15	10	5	1
28	51	48	45	42	39	36	33	28	23	17	12	8	3
29	52	49	46	43	43	37	35	29	24	19	14	10	5
30	53	50	47	44	42	39	33	31	26	21	16	12	7
32	55	52	49	46	44	41	39	34	29	24	20	15	11
34	58	54	51	48	46	43	41	36	32	27	23	19	15
36	58	55	53	50	43	45	43	38	3*	30	26	22	18
38	59	57	54	52	53	47	45	40	36	32	28	24	20
40	61	58	56	53	51	49	47	42	38	34	30	27	23
42	62	60	57	55	53	50	48	44	40	36	32	29	25
44	63	61	58	56	54	52	53	46	42	38	34	31	27
46	64	62	59	57	55	53	51	47	43	39	36	33	29
48	65	63	CO	58	55	54	52	48	4*i	41	37	34	31
50	65	63	61	59	57	55	53	49	45	42	38	36	32

Appendix B

Mean monthly values of extraterrestrial radiation (MJ m^{-2} day^{-1}) for different latitudes in Northern Hemisphere

Month	0°	10°	20°	30°	40°
January	35.94	31.71	26.69	20.04	15.31
February	37.11	34.35	30.50	25.94	20.79
March	37.45	36.57	34.48	31.46	27.49
April	36.36	37.49	37.53	36.40	34.10
May	34.43	37.11	38.83	39.62	39.16
June	33.18	36.61	39.12	40.75	41.30
July	33.60	36.74	39.12	40.04	40.29
August	35.23	37.03	38.91	37.57	36.28
September	36.78	36.61	37.91	33.22	29.96
October	37.03	34.94	35.48	27.74	22.93
November	36.07	32.38	27.82	22.51	16.69
December	35.40	30.84	25.56	19.46	13.77

Maximum possible hours of Sunshine (N_0) for different latitudes in Northern Hemisphere

Month	0°	10°	20°	30°	40°
January	12.1	11.6	11.1	10.4	9.7
February	12.1	11.8	11.5	11.1	10.6
March	12.1	12.1	12.0	12.0	11.9
April	12.1	12.3	12.6	12.9	13.2
May	12.1	12.6	13.1	13.6	14.3
June	12.1	12.7	13.3	14.0	15.0
July	12.1	12.6	12.8	13.9	14.7
August	12.1	12.5	13.2	13.2	13.8
September	12.1	12.2	12.3	12.4	12.5
October	12.1	11.9	11.7	11.5	11.2
November	12.1	11.7	11.2	10.7	10.0
December	12.1	11.6	10.9	10.3	9.4

© Springer International Publishing AG 2017
L. Ahmad et al., *Experimental Agrometeorology: A Practical Manual*,
https://doi.org/10.1007/978-3-319-69185-5

Appendix C

Agro-climatic zones in India (NARP, ICAR)

Abbreviation	Agro-climatic zone
North India	
Jammu and Kashmir	
AZ 1	Low Altitude SubtroPical
AZ 2	Intermediate
AZ 3	Valley Temperate
AZ 4	D Temperate
AZ 5	Cold Arid
Himachal Pradesh	
AZ 6	High Hills Temperate Wet
AZ 7	Submontane And Low Hills Subtropical
AZ 8	Mid Hills Subtropical
AZ 9	Submontane And Low Hills Subtropical
Punjab	
AZ 10	Undulating Plain
AZ 11	Central Plain
AZ 12	Western Plain
AZ 13	Western
AZ 14	Submontaneundulatin
Haryana	
AZ 15	Eastern
AZ 16	Western
Rajasthan	
AZ 17	Arid Western Plain
AZ 18	Irrigated North Western Plain
AZ 19	Transitional Plain Zone Island Drainage
AZ 20	Transitional Plain Zone Of Luni Basin
AZ 21	Semiarid Eastern Plain
AZ 22	Flood Prone Eastern Plain

(continued)

© Springer International Publishing AG 2017
L. Ahmad et al., *Experimental Agrometeorology: A Practical Manual*,
https://doi.org/10.1007/978-3-319-69185-5

(continued)

Abbreviation	Agro-climatic zone
AZ 23	Sub Humid Southern Plain And Alluvial Hill
AZ 24	Southern Humid Plain
AZ 25	South Eastern Humid Plain
Uttaranchal	
AZ 26	Hill
AZ 27	Bhabar and Tarai
Uttar Pradesh	
AZ 28	Western Plain
AZ 29	Mid-Western Plain
AZ 30	South Western Semi-arid
AZ 31	Central Plain
AZ 32	BundelKhand
AZ 33	North Eastern Plain
AZ 34	Eastern Plain
AZ 35	Vindyan
East and North-east India	
West Bengal	
AZ 36	Hilly
AZ 37	Tarai
AZ 38	Old Alluvial
AZ 39	New Alluvial
AZ 40	Laterite And Red Soil Zone
AZ 41	Coastal Saline
Assam	
AZ 42	North Bank Plains
AZ 43	Upper Brahmaputra Valley
AZ 44	Central Brahmaputra Valley
AZ 45	Lower Brahmaputra Valley
AZ 46	Barak Valley
AZ 47	Hills
Arunachal Pradesh	
AZ 48	Alpine
AZ 49	Temperate Sub Alpine
Meghalaya	
AZ 50	Sub-Tropical Hill
Manipur	
AZ 51	Sub-Tropical Plain
Nagaland	
AZ 52	Mid Tropical Hill
Tripura	

(continued)

(continued)

Abbreviation	Agro-climatic zone
AZ 53	Mid Tropical Plain
Bihar and Jharkhand	
AZ 54	Northwest Alluvial Plain
AZ 55	North East Alluvial Plain
AZ 56	South Bihar Alluvial Plain
AZ 57	Central And Northeastern Plateau
AZ 58	Western Plateau
AZ 59	South Eastern Plateau
Orissa	
AZ 60	North Western Plateau
AZ 61	North Central Plateau
AZ 62	North Eastern Coastal Plain
AZ 63	East & Southeastern Coastal Plain
AZ 64	North Eastern Ghat
AZ 65	Eastern Ghat Highland
AZ 66	Southeastern Ghat
AZ 67	Western Undulating
AZ 68	West Central Table
AZ 69	Peninsular India
Peninsular India	
Madhya Pradesh and Chattisgarh	
AZ 70	Chattisgarh Plain Zone Including Chattisgarh Districts
AZ 71	Bastar Plateau
AZ 72	North Hill Zone Of Chattisgarh
AZ 73	Kymora Plateau And Sat ura Hill
AZ 74	Vin a Plateau
AZ 75	Central Narmada Valle
AZ 76	Grid Region
AZ 77	Bundelkhand
AZ 78	Satura Plateau
AZ 79	Malwa Plateau
AZ 80	NimarVall
AZ 81	Jhabua Hills
Gujarat	
AZ 82	East Gujarat heavy rainfall
AZ 83	South Gujarat
AZ 84	Middle Gujarat
AZ 85	North Gujarat
AZ 86	North Western Gujarat
AZ 87	South Saurashtra

(continued)

(continued)

Abbreviation	Agro-climatic zone
AZ 88	North Saurashtra
AZ 89	Ghat and Coastal
Maharashtra	
AZ 90	South Konkan Coastal
AZ 91	North Konkan Coastal
AZ 92	Western Ghat
AZ 93	Submontane
AZ 94	Western Maharashtra Plain
AZ 95	Scarci Zone
AZ 96	Central Maharashtra plateau
AZ 97	Central Vidarbha
AZ 98	Eastern Vidarbha
Karnataka	
AZ 99	North East transition
AZ 100	North east dry
AZ 101	Northern dry
AZ 102	Central dry
AZ 103	Eastern d
AZ 104	Southern dry
AZ 105	Southern transition
AZ 106	Western transition
AZ 107	Hill
AZ 108	Coastal
Kerala	
AZ 109	Northern
AZ 110	Southern
AZ 111	Central
AZ 112	High Altitude
AZ 113	Problem area
Andhra Pradesh	
AZ 114	North Coastal
AZ 115	Southern
AZ 116	Northern Telengana
AZ 117	Scarce rainfall zone of Rayalseema
AZ 118	Southern Telengana
AZ 119	High altitude and tribal
AZ 120	Krishna Godavari
Tamil Nadu	
AZ 121	North eastern
AZ 122	North western

(continued)

(continued)

Abbreviation	Agro-climatic zone
AZ 123	Western
AZ 124	Kavery delta
AZ 125	Southern
AZ 126	High rainfall
AZ 127	High altitude and hilly

Standard meteorological weeks

Week no.	Dates	Week no.	Dates
1	01–07 Jan	27	02–08 Jul
2	08–14 Jan	28	09–15 Jul
3	15–21 Jan	29	16–22 Jul
4	22–28 Jan	30	23–29 Jul
5	29–04 Feb	31	30–05 Aug
6	05–11 Feb	32	06–12 Aug
7	12–18 Feb	33	13–19 Aug
8	19–25 Feb	34	20–26 Aug
9[a]	26 Feb–04 Mar	35	27 Aug–02 Sep
10	05–11 Mar	36	03–09 Sep
11	12–18 Mar	37	10–16 Sep
12	19–25 Mar	38	17–23 Sep
13	26–01 Apr	39	24–30 Sep
14	02–08 Apr	40	01–07 Oct
15	09–15 Apr	41	08–14 Oct
16	16–22 Apr	42	15–21 Oct
17	23–29 Apr	43	22–28 Oct
18	30 Apr–06 May	44	29 Oct–04 Nov
19	07–13 May	45	05–11 Nov
20	14–20 May	46	12–18 Nov
21	21–27 May	47	19–25 Nov
22	28 May–03 Jun	48	26 Nov–02 Dec
23	04–10 Jun	49	03–09 Dec
24	11–17 Jun	50	10–16 Dec
25	18–24 Jun	51	17–23 Dec
26	25 Jun–01 Jul	52[b]	24–31 Dec

[a]Week No. 9 will be 8 days during leap year
[b]Week No. 52 will always have 8 days

References

Ganaie, S.A., Bhat, M.S., Kuchay, N.A. and Parry, J.A. (2014). Delineation of micro agro-climatic zones of jammu and kashmir. *Int. J. Agricult. Stat. Sci.* 10(1): 219–225.

Harrold, L.L. and Dreibelbis, F.R. (1967). Agricultural hydrology as evaluated by monolith lysimeters 1944–1955. *USDA Technical Bulletin*, (1367).

Hoogenboom, G., Jones, J.W., Wilkens, P.W., Porter, C.H., Batchelor, W.D., Hunt, L.A., Boote, K.J., Singh, U., Uryasev, O., Bowen, W.T. and Gijsman, A.J. (2004). Decision support system for agrotechnology transfer version 4.0 [CD-ROM]. University of Hawaii, Honolulu, HI, 1, pp. 235–265.

Kendall, M.G. (1975). Rank Correlation Methods, Fourth edition, Charles Griffin: London.

Krishnan, A. and Singh, M. (1968). January. Soil climatic zones in relation to cropping patterns. In *Proceedings of the Symposium on Cropping Patterns, Indian Council of Agricultural Research, New Delhi* (pp. 172–185).

Khanna, S.S. (1989). The agro-climatic approach. *Survey of Indian agriculture*, pp. 28–35.

Mann, H.B. (1945). Non-parametric test against trend. *Econometrica*, 13, 245–259.

Nuttonson, M.Y. (1948). Some preliminary observations of phenological data as a tool in the study of photoperiodic and thermal requirements of various plant material. Page 129–143 in Vernalization and photoperiodism. A Symposium. Waltham, Mass., Chronica Botanica.

Sastry, P.S.N. and Chakravarty, N.K., (1982). Energy summation indices for wheat crop in India. *Agricultural Meteorology*, 27(1–2), pp. 45–48.

World Meteorological Organization, (2001). Lecture Notes for Training Agricultural Meteorological Personnel (J. Wieringa and J. Lomas). Second edition (WMO-No. 551), Geneva.

World Meteorological Organization, (1992) International Meteorological Vocabulary. Second edition, WMO-No. 182, Geneva.

World Meteorological Organization/United Nations Educational, Scientific and Cultural Organization, (1992). International Glossary of Hydrology. WMO-No. 385, Geneva.

World Meteorological Organization (2003). Manual on the Global Observing System. Vol. I (WMO-No. 544), Geneva.

© Springer International Publishing AG 2017   159

L. Ahmad et al., *Experimental Agrometeorology: A Practical Manual*,

https://doi.org/10.1007/978-3-319-69185-5

Printed in the United States
By Bookmasters